Hybrid Wireless-Power Line Communication for Indoor IoT Networks

For a complete listing of the *Artech House Telecommunications and Network Engineering Series*, turn to the back of this book.

Hybrid Wireless-Power Line Communication for Indoor IoT Networks

Xu Zhu

Kainan Zhu

Mohammad Heggo

ARTECH HOUSE

BOSTON | LONDON
artechhouse.com

Library of Congress Cataloging-in-Publication Data
A catalog record for this book is available from the U.S. Library of Congress.

British Library Cataloguing in Publication Data
A catalogue record for this book is available from the British Library.

Cover design by John Gomes

ISBN 13: 978-1-63081-809-8

685 Canton Street
Norwood, MA 02062

10 9 8 7 6 5 4 3 2 1

Contents

Preface

The Internet of Things (IoT) is one of the fastest-growing technology sectors in the world. Power line communication (PLC), TV white space (TVWS), and wireless sensor networks (WSNs) are regarded as effective and complementary solutions for indoor IoT networks to meet high demands for data rate and coverage. TVWS and WSNs enable flexible and dynamic wireless networking, while PLC utilizes the widespread presence of power line (PL) cables as a communication medium. Hence, their deployment cost is low. However, challenges exist with regard to the coexistence of PLC and wireless communication technologies. This book is dedicated to addressing the challenges involved in hybrid wireless power line communication indoor IoT networks. We hope that the contents of this book will bring inspiration to the readers.

1

Introduction

1.1 Background

The Internet of Things (IoT) envisions common physical objects to be equipped with transceivers and microcontrollers for telecommunications [1], and through properly designed protocol stacks, these objects are enabled to share information and cooperate with each other and with users [2]. With such a communication paradigm [3], the IoT will boost the development of a wide variety of applications, such as home and industrial automation, elderly care, home security surveillance, energy management, and smart grids [4–9], which utilize the tremendous amount and diversity of information gathered from these everyday objects, such as surveillance cameras, home appliances, and monitoring sensors, to provide new services to customers [2].

In the realm of IoT, wireless sensor networks (WSNs) are widely deployed in a plethora of applications due to their advantages such as rapid deployment, low cost, and high flexibility [10,11]. For example, cameras and acoustic sensors are connected to improve the reliability and efficiency of IoT networks in smart grid applications and enable multimedia communication infrastructure [12]. However, WSNs are confronted with a dynamic wireless medium, limited bandwidth, limited energy supply (in battery-powered WSNs), and the often-present blind spot problems [13,14]. In order to tackle these problems, extensive studies [15–24] have been carried out to investigate the performance improvement (e.g., throughput, robustness, energy consumption, network lifetime) on a hybrid network (i.e., mixed wireless and wireline network). The basic idea of these studies is that by adding a few wirelines as the shortcut in the network (termed small-world network [20]), the average hop count can be reduced drastically and thus improve the network performance. However, these studies focus on the analysis of network performance improvement based on a single protocol layer, such as the physical layer or network layer, in which the resources (e.g., bandwidth, energy) of the hybrid network are not utilized efficiently. For example, in order to improve the quality of service (QoS) of the network, both the channel capacity and the bit error rate (BER) at the physical (PHY) layer and the channel access method at the medium access control (MAC) layer, as well as routing, source rate adaptation, and so forth, at the upper layers should be jointly considered. Therefore, a cross-layer design of the hybrid network is necessary. In addition, indoor-battery-powered WSNs are suitable for applications such as chemical and petroleum refining industrial process monitoring (as mains-powered WSNs are prohibited due to safety concerns) and impromptu surveillance installation due to the advantages of discreet and unobtrusive installation and removal. As well, indoor-battery-powered WSNs are immune to the failure of the power distribution system. In such applications, the indoor-battery-powered WSNs often have stringent energy budgets, and so how to prolong the network lifetime becomes a challenging problem.

In order to fulfill the high data rate services in IoT, the demand for radio spectrum grows rapidly. On the other hand, vast measurement data has shown that the scarcity problem in the radio spectrum is mainly caused by inefficient spectrum utilization. For example, the results of spectrum occupancy measurements in [25–29], which were held in different countries, show that the radio spectrum occupancy level does not exceed 25%. This yields more attention toward a cognitive radio solution in high-speed applications such as TV video streaming [30], which allows unlicensed users to access licensed spectrum without causing harmful interference to incumbent licensed users. In May 2004, the Federal Communications Commission (FCC) recommended the cognitive access of the temporally or geographically underutilized channels in the TV spectrum known as TV white space (TVWS) [31]. In November 2008, the FCC released regulatory rules that allow the unlicensed users (known as secondary users (SUs)) to utilize the very high frequency (VHF) band (30 MHz–300 MHz) and the ultrahigh frequency (UHF) band (300 MHz–3 GHz) in the TVWS [32]. The regulations guarantee maximum protection for the primary users (PUs), which can be translated to more spectrum access restrictions on the TV band devices (TVBDs). The TVBDs should acquire a detailed map for the free TVWS spectrum within a certain geographic area and time slot through geolocation database access and/or spectrum sensing techniques. Following the FCC regulation, other international organizations such as the Institute of Electrical and Electronics Engineers (IEEE) [33] and the European Computer Manufacturers Association (ECMA) [34] began the journey to define the PHY and MAC layer standards for TVWS communications.

Power line communication (PLC) is considered a promising technology for IoT data transmission that utilizes the widespread presence of low-voltage (e.g., 220 V) power line (PL) cables as a communication medium [35], and hence offers a cost-effective wired solution. This ubiquitous infrastructure enables every line-powered device to become the potential target of value-added services [36]. PLC is adopted in advanced energy services such as automatic electricity meter reading, supply management, and energy control [37]. Owing to the high data rates that can be supported by PLC (comparable with domestic Ethernet and WiFi), another attractive application of PLC lies in the area of home networking [38–40], in which the advantage of the pervasive presence of PL cables and outlets is fully capitalized [41]. With the recent advances in PLC technology and regulatory and standardization efforts [42], it was announced by chip manufacturers of PLC devices that millions of these devices are being shipped annually for the application of indoor and smart grid communication and the number is expected to grow in the future [43]. However, low-voltage power line cables are not designed for high-speed communications. The broadband signal therefore suffers radiation especially in the VHF band. This can yield electromagnetic interference (EMI) with the existing wireless services [44–46]. To prevent from severe EMI, the transmission power spectral density (PSD) of broadband PLC (BPLC) [47] in the frequency band below 30 MHz is restricted to −55 dBm/Hz and above 30 MHz to −85 dBm/Hz, respectively, according to both the IEEE 1901 [48] and the HomeplugAV2 [49] standards. The strict restriction over the PSD of the BPLC limits the maximum achievable throughput. In other words, the achievable throughput in the permissible VHF band in BPLC (30 MHz–100 MHz) is limited compared to the HF band (2 MHz–30 MHz), which implies bandwidth efficiency loss. This represents a crucial problem in BPLC, which requires an innovative solution for better exploitation of the BPLC VHF band.

As wireless communications (WSNs and TVWS) and PLC technologies are the building blocks of the IoT [3] and supplementary to each other, the integration of existing communication technologies is the trend for future communication networks [16]. This trend of the flourishing of hybrid networks can be forecast from the recent advances in the development of software-defined networking (SDN) [50–52]. An architecture design of using PLC as a backhaul for WSNs to implement a smart home control network is reported in [53]. Recent studies in [54–64] have reported a performance improvement in the indoor home networks or smart grids by exploiting the diversity of hybrid wireless and PLC channels. These studies either utilize the hybrid channel in serial or in parallel to improve the network throughput or reliability. Nevertheless, similar to the work in small-world networks [15–24], these studies analyzed the performance of such a hybrid channel in an information theoretic framework based on a single protocol layer, by which the resources in the hybrid channel are not utilized efficiently. Therefore, it is necessary to investigate the performance improvement of such a hybrid network.

1.2 Book Organization

The rest of this book contains three parts. Part I (Chapters 2–3) introduces the fundamentals of wireless and power line communications. In Chapter 2, an overview of the wireless and PLC standards, channel characteristics, and challenges is presented. Chapter 3 provides details on the channel models of wireless communications and PLC. Part II of this book (Chapters 4–7) studies the hybrid VHF TVWS-BPLC for indoor high-speed IoT networks, with the aim of complementing with each other to deliver enhanced performance. In Chapter 4, an overview and applications of IoT networks are provided. Chapter 5 studies the crosstalk between TV white space and PLC. In Chapter 6, ergodic capacity is investigated for a hybrid TVWS BPLC point-to-point system. In Chapter 7, WBPLC point-to-multipoint system is proposed, where the MAC and the PHY layers are developed for the new system in order to be TVWS standard compliant. The main original contributions of this part are described as follows.

In Chapter 5, a general statistical based path loss mapping (GSBPL) approach is proposed for modeling the path loss of indoor low voltage (i.e., 220 V) BPLC. As well, a simplification method is proposed for computing the channel transfer function, which is proven to be more general and computationally more efficient than the previous method in literature. The feasibility of the cooperation between BPLC and wireless communications is thus concluded by comparing their corresponding path losses.

In Chapters 6 and 7, new hybrid systems utilizing BPLC and TVWS are proposed in the VHF band referred to as white BPLC (WBPLC). Two cases are considered in the proposed system: (1) a point-to-point WBPLC multiple-input multiple-output (MIMO) system, where a power allocation algorithm and an iterative precoding technique are proposed to maximize the ergodic capacity subject to the constraints of total power and interference limit at the TV (PU receiver (Rx), and (2) a point-to-multipoint WBPLC MIMO system. The overall network downlink capacity maximization problem is investigated using an efficient algorithm for power and subcarrier allocation among different users.

Part III of this book (Chapters 9–10) studies hybrid UHF wireless-PL sensor networks. Chapter 8 provides an overview and the applications of sensor networks through cross-layer design. A joint optimal design of PHY, MAC, and network layers to maximize the network lifetime for hybrid wireless-PL sensor networks is presented in Chapter 9. In Chapter 10, a cross-layer video encoding, channel access control, and link rate allocation scheme is developed for hybrid wireless-PL video sensor networks. The main original contributions of this part are as follows.

In Chapter 9, a hybrid sensor network for industrial sensor network applications, which consists of both wireless and PL sensor nodes, is proposed.

Since the data rate requirement for such applications is typically low, and in the ease of derivation, the power consumption model takes into account the transmission signal power and the power consumption of the power amplifier. First, to the best of our knowledge, it is the first reported work in the literature that focuses on the cross-layer design of such a heterogeneous network. The hybrid sensor network takes the advantage of the flexibility of WSNs while the PL sensors are deployed to prolong the lifetime of the network. This work studies the joint design of the PHY, MAC, and network layers to maximize the hybrid network lifetime, which is limited by the battery capacity of wireless sensors. Second, closed-form expressions of the globally optimal solution for lifetime maximization of the hybrid sensor network are derived for the linear topology. Such closed-form solutions give insights into factors that are significant to the network lifetime when designing the hybrid sensor network. Third, the impacts of different network configurations, such as source rate and sensor node densities on the hybrid network lifetime, are investigated. The impact of different transmission strategies of PL nodes on the effectiveness of the network is studied.

In Chapter 10, a hybrid video sensor network (HVSN), which comprises both battery-powered wireless sensor nodes and PL sensor nodes, is proposed to maximize the network lifetime. Since HVSNS have a high data rate requirement, the power consumption model includes the power consumption due to video encoding, data transmission, and reception. First, to the best of our knowledge, it is the first reported work to investigate video sensor networks with hybrid power sources and hybrid communication schemes. The proposed HVSN utilizes the flexibility of wireless nodes while PL nodes are used to extend the network lifetime. Second, the joint design of video encoding rate, aggregate power consumption, and channel access control, along with link rate allocation, is studied for maximizing the hybrid network lifetime. The joint design achieves much better performance than separate optimization. Third, a distributed algorithm is proposed for the network lifetime maximization problem. The distributed algorithm divides the computational burden among all nodes with much lower communication overhead. Fourth, the impact of dynamic network change and network scalability is studied. The effectiveness of the proposed algorithm is validated through extensive simulation results.

References

[1] Atzori, L., A. Iera, and G. Morabito, The Internet of Things: A Survey, *Computer Networks*, Vol. 54, No. 15, 2010, pp. 2787–2805.

[2] Zanella, A., N. Bui, A. Castellani, L. Vangelista, and M. Zorzi, "Internet of Things for Smart Cities," *IEEE Internet of Things Journal*, Vol. 1, No. 1, 2014, pp. 22–32.

[3] Yaqoob, I., E. Ahmed, I. Abaker, et al., "Internet of Things Architecture: Recent Advances, Taxonomy, Requirements, and Open Challenges," *IEEE Wireless Communications*, Vol. 24, No. 3, 2017, pp. 10–16.

[4] Fadel, E., V. C. Gungor, L. Nassef, et al., "A Survey on Wireless Sensor Networks for Smart Grid," *Computer Communications*, Vol. 71, No., 2015, pp. 22–33.

[5] Bellavista, P., G. Cardone, A. Corradi, and L. Foschini, "Convergence of MANET and WSN in Iot Urban Scenarios," *IEEE Sensors Journal*, Vol. 13, No. 10, 2013, pp. 3558–3567.

[6] Hu, L., M.Qiu, J. Song, M. S. Hossain, and A. Ghoneim, "Software Defined Healthcare Networks," *IEEE Wireless Communications*, Vol. 22, No. 6, 2015, pp. 67–75.

[7] Samanta, A., S. Bera, and S. Misra, "Link-Quality-Aware Resource Allocation with Load Balance in Wireless Body Area Networks," *IEEE Systems Journal*, Vol. 12, No. 1, 2015, pp. 74–81.

[8] Pan, J., R. Jain, S. Paul, T. Vu, A. Saifullah, and M. Sha, "An Internet of Things Framework for Smart Energy in Buildings: Designs, Prototype, and Experiments," *IEEE Internet of Things Journal*, Vol. 2, No. 6, 2015, pp. 527–537.

[9] Bera, S., S. Misra, and J. JPC Rodrigues. "Cloud Computing Applications for Smart Grid: A Survey," *IEEE Transactions on Parallel and Distributed Systems*, Vol. 26, No. 5, 2015, pp. 1477–1494.

[10] Sitharama Iyengar, S., and R. R. Brooks, *Distributed Sensor Networks: Sensor Networking and Applications*, Boca Raton, FL: CRC Press, 2016.

[11] Al-Fuqaha, A., M. Guizani, M. Mohammadi, M. Aledhari, and M. Ayyash, "Internet of Things: A Survey on Enabling Technologies, Protocols, and Applications," *IEEE Communications Surveys & Tutorials*, Vol. 17, No. 4, 2015, pp. 2347–2376.

[12] Wang, H., Y. Qian, and H. Sharif, "Multimedia Communications over Cognitive Radio Networks for Smart Grid Applications," *IEEE Wireless Communications*, Vol. 20, No. 4, August 2013, pp. 125–132.

[13] Akyildiz, I. F., W. Su, Y. Sankarasubramaniam, and E. Cayirci, "Wireless Sensor Networks: A Survey," *Computer Networks*, Vol. 38, No. 4, 2002, pp. 393–422.

[14] Alduraibi, F., N. Lasla, and M. Younis, "Coverage-Based Node Placement Optimization in Wireless Sensor Network with Linear Topology," in *Proc. IEEE International Conference on Communications (ICC)*, Kuala Lumpur, Malaysia, May 2016, pp. 1–6.

[15] Sharma G., and R. R. Mazumdar," A Case for Hybrid Sensor Networks," *IEEE/ACM Transactions on Networking*, Vol. 16, No. 5, 2008, pp. 1121–1132.

[16] Chen, H.-Y., and C.-H. Lee, "A Spatial Study of Mixed Wireless and Wireline Heterogeneous Networks," in *Proc. IEEE International Symposium on Personal Indoor and Mobile Radio Communications (PIMRC)*, Toronto, Canada, 2011, pp. 258–262.

[17] Salvadori, F., C. S. Gehrke, A. C. De Oliveira, M. De Campos, and P. Sergio Sausen, "Smart Grid Infrastructure Using a Hybrid Network Architecture," *IEEE Transactions on Smart Grid*, Vol. 4, No. 3, 2013, pp. 1630–1639.

[18] Qiu, T., D. Luo, F. Xia, N. Deonauth, W. Si, and A. Tolba, "A Greedy Model with Small World for Improving the Robustness of Heterogeneous Internet of Things," *Computer Networks*, Vol. 101, 2016. pp. 127–143.

[19] Helmy, Ah., "Small Worlds in Wireless Networks," *IEEE Communications Letters*, Vol. 7, No. 10, 2003, pp. 490–492.

[20] Sharma, G., and R. Mazumdar, "Hybrid Sensor Networks: A Small World," in *Proc. ACM International Symposium on Mobile Ad Hoc Networking and Computing*, Chicago, Illinois, 2005, pp. 366–377.

[21] Dongyang, W., W.Muqing, L. Bo, and W. Jingrong," A Small World-Based Energy-Efficient Mechanism in Wireless Ad Hoc Networks," in *Proc. IEEE International Conference on Communications (ICC)*, Budapest, Hungary, 2013, pp. 447–451.

[22] Bo, L., W. Muqing, W. Jingrong, and W. Dongyang," Small Worlds in Multi-Channel Wireless Networks: An Analytical Approach," *in Proc. IEEE International Conference on Communications (ICC)*, Budapest, Hungary, 2013, pp. 1527–1531.

[23] Guidoni, D. L., R. A. F. Mini, and A. A. F. Loureiro. "Applying the Small World Concepts in the Design of Heterogeneous Wireless Sensor Networks," *IEEE Communications Letters*, Vol. 16, No. 7, 2012, pp. 953–955.

[24] Tadayon, N., A. E. Zonouz, S. Aıssa, and L. Xing," Cost Effective Design and Evaluation of Wireless Sensor Networks Using Topology Planning Methods in Small-World Context," *IET Wireless Sensor Systems*, Vol. 4, No. 2, 2014, pp. 43–53.

[25] Patil, K., R. Prasad, and K. Skouby, "A Survey of Worldwide Spectrum Occupancy Measurement Campaigns for Cognitive Radio," in *Proc. International Conference on Devices and Communications (ICDeCom)*, February 2011, Mesra, India, pp. 1–5.

[26] Contreras, S., G. Villardi, R. Funada, and H. Harada, "An Investigation into the Spectrum Occupancy in Japan in the Context of TV White Space Systems," in *6th International ICST Conference on Cognitive Radio Oriented Wireless Networks and Communications (CROWNCOM)*, June 2011, Osaka, Japan, pp. 341–345.

[27] Martian, A., C. Vladeanu, I. Marcu, and I. Marghescu, "Evaluation of Spectrum Occupancy in an Urban Environment in a Cognitive Radio Context," *International Journal on Advances in Telecommunications*, Vol. 3, No. 3, 2010, pp. 172–181.

[28] Hoyhtya, M., A. Mammela, M. Eskola, et al., "Spectrum Occupancy Measurements: A Survey and Use of Interference Maps," *IEEE Communications Surveys and Tutorials*, Vol. 99, April 2016, pp. 1–30.

[29] Islam, M. H., C. L. Koh, S. W. Oh, et al., "Spectrum Survey in Singapore: Occupancy Measurements and Analyses," in *3rd International Conference on Cognitive Radio Oriented Wireless Networks and Communications (CrownCom 2008)*, May 2008, Singapore, pp. 1–7.

[30] Fadda, M., M. Murroni, and V. Popescu, "A Cognitive Radio Indoor HDTV Multi-Vision System in the TV White Spaces," *IEEE Transactions on Consumer Electronics*, Vol. 58, No. 2, May 2012, pp. 302–310.

[31] Federal Communications Commission, Notice of Proposed Rule Making, Document 04–113, May 2004.

[32] Federal Communication commission, Second Report and Order and Memorandum Opinion and Order in the Matter of Unlicensed Operation in the TV Broadcast Bands, Additional Spectrum for Unlicensed Devices Below 900 MHz and in the 3 GHz Band, Document 08-260, November 2008.

[33] IEEE 802.11 af Draft 5.0, Amendment 5: TV White Spaces Operation, Retrieved April 2013.

[34] MAC and PHY for Operation in TV White Space, ECMA-392, December 2009.

[35] Ferreira, H. C., L. Lampe, J. Newbury, T. Swart (eds.), *Power Line Communications: Theory and Applications for Narrowband and Broadband Communications over Power Lines*, Chichester, United Kingdom: John Wiley & Sons, 2010.

[36] Galli, S., and O. Logvinov, "Recent Developments in the Standardization of Power Line Communications within the IEEE," *IEEE Communications Magazine*, Vol. 46, No. 7, 2008.

[37] Pavlidou, N., A. J. Han Vinck, J. Yazdani, and B. Honary, "Power Line Communications: State of the Art and Future Trends," *IEEE Communications Magazine*, Vol. 41, No. 4, 2003, pp. 34–40.

[38] Lin, C.-K., H.-W. Chu, S.-C. Yeh, M.-T. Lu, J. Yao, and H.r Chen, "Robust Video Streaming over Power Lines," in *Proc. IEEE International Symposium on Power Line Communications and Its Applications (ISPLC)*, Orlando, Florida, 2006, pp. 196–201.

[39] Gnazzo, A., A. Bergaglio, M. Palma, F. Pittoni, M. Giunta, and F. Ballesio, "Powerline Technology over Coaxial Cables for In-Home Multimedia Applications: Performances and EMC Issues," in *Proc. IEEE International Symposium on Power Line Communications and Its Applications (ISPLC)*, Udine, Italy, 2011, pp. 130–134.

[40] Lampe, L., *Power Line Communications: Principles, Standards and Applications from Multimedia to Smart Grid*, Chichester, United Kingdom: John Wiley & Sons, 2016.

[41] Galli, St., and T. C. Banwell, "A Deterministic Frequency-Domain Model for the Indoor Power Line Transfer Function," *IEEE Journal on Selected Areas in Communications*, Vol. 24, No. 7, 2006, pp. 1304–1316.

[42] Held, G., *Understanding Broadband over Power Line*, Boca Raton, FL: CRC Press, 2016.

[43] Cano, C., A. Pittolo, D. Malone, L. Lampe, A. M. Tonello, and A. G. Dabak, "State of the Art in Power Line Communications: From the Applications to the Medium," *IEEE Journal on Selected Areas in Communications*, Vol. 34, No. 7, 2016, pp. 1935–1952.

[44] Wang, L. B., P. L. So, K. Y. See, M. Oswal, and T. S. Pang, "Investigation of Radiated Emissions in Power Line Communication Networks," in *Proc. IEEE International Power Engineering Conference (IPEC 2007)*, December 2007, Singapore, pp. 455–460.

[45] Khedimallah, S., B. Nekhoul, K. Kerroum, and K. El Khamlichi Drissi, "Analysis of Power Line Communications Electromagnetic Field in Electrical Networks Taking into Account the Power Transformers," in *Proc. IEEE International Symposium on Electromagnetic Compatibility*, Septemer 2012, Rome, Italy, pp. 1–6.

[46] Doric, V., D. Poljak, I. Hadjina, and K. El Khamlichi Drissi, "EMC Analysis of the Narrowband PLC System Based on the Antenna Theory," in *Proc. IEEE 21st Int. Conf. Soft., Telecomm. Comput. Networks (SoftCOM)*, September 2013, Primosten, Croatia, pp. 1–5.

[47] Rehman, M. U., S. Wang, Y. Liu, S. Chen, X. Chen, and C. G. Parini, "Achieving High Data Rate in Multiband-OFDM UWB over Power-Line Communication System," *IEEE Transactions on Power Delivery*, Vol. 27, No. 3, July 20102, pp. 172–1177.

[48] IEEE Standards Association, et al., IEEE Standard for Broadband over Power Line Networks: Medium Access Control and Physical Layer Specifications. IEEE Std 1901, 2010:1–1586, 2010.

[49] Berger, L. T., A. Schwager, P. Pagani, and D. Schneider, *MIMO Power Line Communications: Narrow and Broadband Standards, EMC, and Advanced Processing*, Boca Raton, FL: CRC Press, 2014.

[50] Nunes, B. A. A., M. Mendonca, X.-N. Nguyen, K. Obraczka, and T. Turletti, "A Survey of Software-Defined Networking: Past, Present, and Future of Programmable Networks," *IEEE Communications Surveys & Tutorials*, Vol. 16, No. 3, 2014, pp. 1617–1634.

[51] Macedo, D. F., D. Guedes, L. F. M. Vieira, M. A. M. Vieira, and M. Nogueira, "Programmable Networks: From Software-Defined Radio to Software-Defined Networking," *IEEE Communications Surveys & Tutorials*, Vol. 17, No. 2, 2015, pp. 1102–1125.

[52] Bera, S., S. Misra, and A. V. Vasilakos, "Software-Defined Networking for Internet of Things: A Survey," *IEEE Internet of Things Journal*, Vol. 4, No. 6, 2017, pp. 1994–2008.

[53] Li, M., and H.-J. Lin, "Design and Implementation of Smart Home Control Systems Based on Wireless Sensor Networks and Power Line Communications," *IEEE Transactions on Industrial Electronics*, Vol. 62, No. 7, 2015, pp. 4430–4442.

[54] Lee J.-H., and Y.-H. Kim, "Diversity Relaying for Parallel Use of Powerline and Wireless Communication Networks," *IEEE Transactions on Power Delivery*, Vol. 29, No. 3, 2014, pp. 1301–1310.

[55] Lai, S. W., and G. G. Messier, "Using the Wireless and PLC channels for Diversity," *IEEE Transactions on Communications*, Vol. 60, No. 12, 2012, pp. 3865–3875.

[56] Lai, S. W., N. Shabehpour, G. G. Messier, and L. Lampe, "Performance of Wireless/Power Line Media Diversity in the Office Environment, in *Proc. IEEE Global Communications Conference* (GLOBECOM), December 2014, Texas, pp. 2972–2976.

[57] Sayed, M., T. A. Tsiftsis, and N. Al-Dhahir, "On the Diversity of Hybrid Narrowband-PLC/Wireless Communications for Smart Grids," *IEEE Transactions on Wireless Communications*, Vol. 16, No. 7, 2017, pp. 4344–4360.

[58] Sebaali, G., and B. L. Evans, "Design Tradeoffs in Joint Powerline and Wireless Transmission for Smart Grid Communications," in *Proc. IEEE International Symposium on Power Line Communications and its Applications (ISPLC)*, Austin, Texas, 2015, pp. 83–88.

[59] Sayed, M., G. Sebaali, B. L. Evans, and N. Al-Dhahir, "Efficient Diversity Technique for Hybrid Narrowband-Powerline/Wireless Smart Grid Communications," in *Proc. IEEE International Conference on Smart Grid Communications (SmartGridComm)*, Miami, FL, 2015, pp. 1–6.

[60] de MBA Dib, L., V. Fernandes, M. de L. Filomeno, and M. V. Ribeiro, "Hybrid Plc/Wireless Communication for Smart Grids and Internet of Things Applications, *IEEE Internet of Things Journal*, Vol. 5, No. 2, 2017, pp. 655–667.

[61] Mokhtar, M., W. U. Bajwa, and N. Al-Dhahir, "Sparsityaware Joint Narrowband Interference and Impulse Noise Mitigation for Hybrid Powerline-Wireless Transmission," in *Proc. IEEE Wireless Communications and Networking Conference* (WCNC), New Orleans, LA, 2015, pp. 615–620.

[62] Mathur, A., M. R. Bhatnagar, and B. K. Panigrahi, "Performance of a Dual-Hop Wireless-Powerline Mixed Cooperative System," in *Proc. IEEE International Conference on Advanced Technologies for Communications* (ATC), Hanoi, Vietnam, 2016, pp. 401–406.

[63] Kuhn, M., S. Berger, I. Hammerstrom, and A. Wittneben, "Power Line Enhanced Cooperative Wireless Communications," *IEEE Journal on Selected Areas in Communications*, Vol. 24, No. 7, 2006, pp. 1401–1410.

[64] Oliveira, T., F. Andrade, A. Picorone, H. Latchman, S. Netto, and M. Ribeiro, "Characterization of Hybrid Communication Channel in Indoor Scenario," *Journal of Communication and Information Systems*, Vol. 31, No. 1, 2016.

Part I:

Fundamentals of Wireless and Power Line Communications

2

Overview of Wireless and Power Line Communications

2.1 Introduction

An overview of the main regulations and standards of both TVWS and PLC technologies is presented. Also, the channel characteristics of the wireless TVWS and the BPLC in the VHF band are shown along with the cross talk between the two channels. The aim of the presentation is to highlight the advantages and the limitations of each technology, and at the same time highlight the fact that the advantages of one technology may be used to overcome the limitations in the other one. As well, previous work in enhancing the capacity of the BPLC channel in the VHF band is presented. Consequently, this drives the motivation to discuss the expected advantages of the cooperation between the TVWS and the BPLC communications in the indoor environment, which exploits the VHF band in TVWS by offering a MIMO solution using both the TVWS wireless channel and the BPLC channel.

2.2 Overview of Wireless Communications

2.2.1 A Brief History of Wireless Communications

The very first stage of wireless communications dates back to the preindustrial era [1], when signal combinations such as torch signaling, smoke signaling, and flag signaling were designed to enable line-of-sight (LoS) information transmission. An example of such a wireless communication system is the signaling of smoke at beacon towers along the Great Wall to warn of enemy attacks in ancient China.

In 1864, the electromagnetic theory of light was developed by James Clerk Maxwell, who also predicted the existence of radio waves [2]. Later in 1887, Heinrich Hertz proved experimentally the physical existence of these radio waves. Based on the pioneering work of Maxwell and Hertz, the field of radio communications was initiated. The first wireless communication system [1] occurred in 1894, when Oliver Lodge managed to send a radio signal at a distance of 150 m. In 1897, the entrepreneur Guglielmo Marconi founded The Wireless Telegraph and Signal Company in the United Kingdom and demonstrated a series of experiments on wireless communications. Years later, Marconi's wireless system became the world's first wireless system that could enable transatlantic communication.

Currently, based on various application scenarios, wireless communication systems have been developed into many types, such as cellular telephone systems, wireless local area networks (WLANs), satellite networks, and Bluetooth, to name a few. Detailed descriptions of the aforementioned application scenarios of wireless communication systems can be found in [1].

2.2.2 TVWS Regulations and Standards

2.2.2.1 Governing Regulations

According to the FCC rules issued in 2008 [3], TVBDs are divided into two categories: (1) fixed devices, and (2) portable/personal devices. Fixed devices are allowed to access a free TV channel with total transmission power of 4 W as effective isotropic radiated power (EIRP). However, for the portable/personal devices, the maximum permissible EIRP is 100 mW and 40 mW for a free nonadjacent and adjacent channel next to an occupied TV channel, respectively. The fixed devices category should have an access over a geolocation database to obtain a list of the free channels and also should have a high sensing capability to detect PU presence even if the PU received signal power is as low as −114 dBm. However, the portable/personal devices category is further subdivided into two subcategories: (1) mode I, which should have an access over a geolocation database, and (2) mode II, which is unnecessary to have a connection with a geolocation database but must have a high sensing capability.

2.2.2.2 Standards

The TVWS standards are developed mainly by two organizations: IEEE [4] and ECMA [5]. IEEE developed five standards to utilize the TVWS spectrum in different applications, which can be summarized as (1) IEEE 802.11af for WLAN, (2) IEEE 802.22 for wireless regional area network (WRAN), (3) IEEE 802.15.4m for low-rate wireless personal area network (LR-WPAN), (4) IEEE 802.19.1 for enabling the family of IEEE 802 wireless standards to effectively utilize the TVWS spectrum by providing coexistence methods among

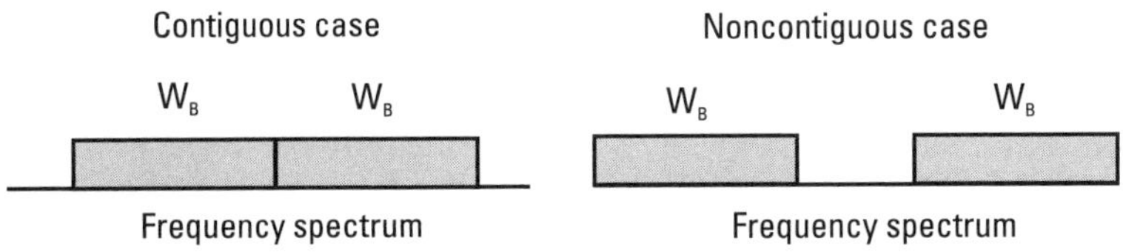

Figure 2.1 Contiguous and noncontiguous TV channels in IEEE 802.11af [4].

Table 2.1
IEEE 802.11af PHY Frame [4]

L-STF	L-LTF	L-SIG	TVHT-SIG-A	TVHT-STF	TVHT-LTF	TVHT-SIG-B	Data Field
8 μs	8 μs	4 μs	4 μs	8 μs	8 μs	4 μs	Data Length

independent TVBD networks, and (5) IEEE 1900.7 for radio interface working in the TVWS spectrum. ECMA also released the standard ECMA-392 first edition in 2009 and the second edition in 2012 for supporting TVWS access. In this chapter, the WLAN applications are the main interesting topic and therefore the IEEE 802.11af and ECMA-392 standards will be discussed in detail.

2.2.2.3 IEEE 802.11af

In December 2013, IEEE 802.11af standard was released to provide the PHY and MAC regulations to organize spectrum sharing among licensed and unlicensed users in the TVWS. The IEEE 802.11af is considered as an amended version of the main standard IEEE 802.11 and its very high throughput (VHT) version of IEEE 802.11ac operating in bands below 6-GHz band. In IEEE 802.11af, two modes are presented: (1) non-high throughput (non-HT), and (2) TV very high throughput (TVHT). The channel bandwidth for non-HT mode is the basic channel unit (BCU) W_B. The BCU is defined as the original bandwidth of the TV channel according to the regulatory domain (i.e., 6 MHz, 7 MHz and 8 MHz). However, the TVHT supports two modes for the channel bandwidth: (1) contiguous mode (i.e., $2W_B$, $4W_B$), where the channel bandwidth spans two or more adjacent BCUs, and (2) noncontiguous mode ($W_B + W_B$, $2W_B + 2W_B$), where the channel bandwidth spans two or more nonadjacent BCUs. Also, the TVHT supports two or more spatial streams, which significantly enhances the achieved throughput. For each orthogonal frequency division multiplexed (OFDM) symbol, 144 subcarriers are used in 6-MHz or 8-MHz channels. However, for 7 MHz, 168 subcarriers are used to maintain the same subchannel bandwidth of the 6-MHz channel. This yields a subcarrier frequency spacing of 41.66 kHz for the 6- and 7-MHz channels, while for the 8-MHz channel, the subcarrier frequency spacing is 55.55 kHz. The duration of the OFDM symbol is 24 μs for the 6- and 7-MHz channels, while it is 18 μs for 8-MHz channel.

Table 2.2
ECMA-392 PHY Frame [5]

Short Preamble	Long Preamble	Header	Data	Tail bits	Pad Bits
1 OFDM symbol	2 OFDM symbol	2 OFDM symbol	1 - 4095 bytes	6 bits	variable

Also the guard interval (GI) period is defined as a quarter of the OFDM symbol duration.

The general frame format for the PHY protocol data unit (PPDU) consists of 8 fields: non-TVHT short training field (L-STF = 8 μs), non-TVHT long training field (L-LTF = 8 μs), non-TVHT signal field (L-SIG = 4 μs), TVHT signal A field (TVHT-SIG-A = 4 μs), TVHT short training field (TVHT-STF = 8 μs), TVHT long training field (TVHT-LTF = 8 μs), TVHT signal B field (TVHT-SIG-B = 4 μs) and data field. The TVHT fields of the general frame exist only in the TVHT mode. The non-TVHT STF is used to improve the automatic gain control (AGC) of the single-input single output (SISO) channel, diversity selection, time acquisition, and coarse frequency estimation at the Rx side. Also, the non-TVHT LTF is used for the channel estimation and frequency acquisition at the receiver (Rx). However, TVHT-STF and TVHT-LTF aim to improve the AGC and channel estimation for the MIMO Rx. The non-TVHT signal field contains the frame header, which holds information about the frame length, rate, parity bits, and tail bits. The TVHT-SIG-A field contains information related to interpreting the TVHT frame, such as channel bandwidth, space time block coding (STBC), and beamforming. The TVHT-SIG-B field contains information about the padding and tail bits in the TVHT frame.

2.2.2.4 ECMA-392

In December 2009, ECMA released its first edition of the ECMA-392 standard to provide the PHY and MAC layers for the TVWS communications. The ECMA-392 standard [5] adopts a channel bandwidth equivalent to the BCU bandwidth (i.e., 6 MHz, 7 MHz, and 8 MHz). Unlike the IEEE 802.11af, the ECMA 392 standard does not support the contiguous or the noncontiguous modes of operation, which means that the channel bandwidth is limited only to one BCU channel. Also, the standard supports using multiple spatial data streams like the TVHT in the IEEE 802.11af. Each TV channel is spanned by 128 subcarriers, with subcarrier frequency spacing of 53.5 kHz, 62.5 kHz, and 71.42 kHz for channel bandwidths of 6 MHz, 7 MHz, and 8 MHz, respectively. The OFDM symbol duration is taken as 18.667 μs, 16 μs, and 13.8 μs for the previously mentioned channel bandwidths. The GI period can be taken as the symbol time duration divided by 8, 16, or 32.

The PHY frame in ECMA-392 is divided into three main fields: (1) preamble, (2) header, and (3) data. The preamble field consists of two subfields: (a) short preamble, which has the duration of one OFDM symbol and is used for automatic gain control (AGC) at the receiver (Rx) side, and (b) long preamble, which has the duration of two OFDM symbols and is used for channel estimation. The header field consists of PHY header, MAC header, parity bits, and tail bits. The PHY header holds the information of data rate, length, scrambler initialisation seed, interleaver option, multiple antenna mode, and transmission power. The MAC header incorporates the information directly from the MAC layer without change. The overall duration of the header is 2 OFDM symbols. The data field consists of the data bytes requested to be sent (maximum 4,095 bytes), the tail bits, and padding bits.

2.2.2.5 MIMO TVWS

MIMO techniques can significantly enhance the spectral efficiency of wireless communication systems in communication channels, which are characterized by multipath propagation due to scattering on different obstacles. Different MIMO techniques, such as space time block coding (STBC) and spatial multiplexing (SM), are allowed in both IEEE 802.11af [4] and ECMA-392 [5]. However, in [6] and [7] two main challenges were presented to using multiple antennas in the TVWS. The first one is due to the long wavelengths in the UHF and VHF bands. The typical wavelength in UHF ranges from 0.1m to 1m, which makes it not practical to design small footprints for MIMO TVBDs. In the VHF, the case is even worse as the wavelength ranges from 1m to 10m. The second challenge is the mutual coupling between multiple antennas that increases in the low frequencies. Hence, most of the MIMO solutions were proposed in the UHF band as in [8], which limits the exploitation of the VHF band of the TVWS.

2.2.2.6 TVWS Channel and Its Path Loss

The TVWS channel differs according to the frequency band of operation. However, the VHF channel characteristics are better than the UHF in terms of path loss and penetration through the concrete walls. More focus will be on the VHF TVWS channel, since it is the shared channel between TVWS and BPLC. According to [9], the path loss of the TVWS is dependent on several factors such as number of walls, number of floors between the Tx and the Rx, in addition to the frequency and separation distance. For example, the average path loss at 100 MHz ranges between 51 dB and 81 dB for a coverage distance between 15m and 50m, respectively. Also, the root mean square (RMS) delay of the TVWS channel is 100 ns, which resembles to 90% coherence bandwidth of 200 kHz [10].

The TVWS channel operating in the VHF band in the indoor environment is modeled in [11] using the general urban path loss model (GUPL). The path loss as a function of the coverage distance can be expressed as [11]:

$$\begin{aligned} GUPL(dB) = &- 10\log_{10}\left(\left[\frac{\lambda}{4\pi d_0}\right]^{\beta}\right) \\ &+ 10n\log_{10}\left(\frac{d}{d_0}\right) + \alpha d + FAF \end{aligned} \tag{2.1}$$

where d is the Tx-Rx geometrical separation, d_0 is the close-in reference distance and it must be chosen to lie in the far field region, which is defined as the region where the radiation pattern does not change shape with distance (although the fields still die off as $1/d$, the power density dies off as $1/d^2$). Also, this region is dominated by radiated fields, with the E- and H-fields orthogonal to each other and the direction of propagation as with plane waves (i.e., $d_0 \gg \frac{2D^2}{\lambda}$). λ is the wavelength, D is the maximum linear dimension of the antenna, β is the power exponent, n is the path loss exponent, α is the attenuation constant (dB/m), and FAF is the floor attenuation factor.

2.3 Overview of Power Line Communications

2.3.1 A Brief Historical Evolution

The very early prototype of PLC dates back to 1838 [12], when remote electricity supply metering was proposed aiming to measure the voltage levels of batteries at unmanned sites. Later in 1897 and 1905, the PL signaling electricity meter was patented in the United Kingdom and United States, respectively [13]. Transmitting voice messages over high-voltage PL occurred in the 1920s [13], when power companies used PL as a substitute for telephone lines (running parallel to the PL) for the communication between transformer stations. Since the telephone circuits are expensive and telephone lines are fragile to harsh environments (e.g., mountainous terrain, bad weather), in 1950, ripple control systems, recognized as the first PLC systems, were designed and then deployed over medium- and low-voltage electrical networks to enable power companies to deliver electricity alongside with commands such as load control and tariff switching to end users [14]. Typically, the aforementioned applications of PLC require low data rates and operate in the narrowband (NB) frequency range of PLC, termed NB PLC, and real-time communication is not required.

With the evolution of technology, the applications in the sense of a modern home network include but are not restricted to simultaneous internet access, shared printers, home control, and remote monitoring, to name a few. In these applications, traditional NB PLC fails to deliver the upsurging data rate requirements. On the other hand, these increasing end-user demands coupled

Table 2.3
Categorization of PLC Technology Based on Operating Bandwidth [15]

PLC Classes	UNB	NB	BB
Bandwidth	0.3–3 kHz or 30–300 Hz	3–500 kHz	1.8–250 MHz
Supported data rate	Around 100 bps	Few kbps to 500 kbps	Several Mbps to several hundred Mbps
Communication range	Up to 150 km [16]	Up to several km	Up to several hundred meters

with the deregulated telecommunication market over the past few decades has fostered the development of what is known as the broadband (BB) PLC. BB PLC operates in high-frequency bands and can achieve very high data rates (several Mbps to several hundred Mbps) while in the meantime is able to provide real-time communication, as required in standard-definition television (SDTV) or high-definition television (HDTV).

PLC technologies can be categorized into three classes according to their operational bandwidth [15]: namely, ultra narrowband (UNB), narrowband (NB), and broadband (BB) PLC, which are specified in standards of IEEE 1901.2 and IEEE 1901. Table 2.3 summarizes the corresponding bandwidth, achievable data rates, and the communication ranges for different PLC technologies.

UNB PLC provides a rather modest data rate, near 100 bits-per-second (bps), while it supports a very large operational range (up to 150 km) mainly due to the fact that it has a small path loss effect (around 0.01 dB/km) and it is able to pass through transformers easily [15]. Such a mature PLC technology has been used in field for at least two decades and it has been deployed for various of utilities. NB PLC support higher data rate than UNB PLC at the expense of transmission range. NB PLC can be further divided into low data rate NB PLC and high data rate NB PLC depending on whether a single carrier or multiple carriers are used. Finally, BB PLC supports data rate up to several hundred Mbps. It is normally limited for the use of home networking due to the fact that it is difficult to pass through transformers and PL signals on different phases may not be able to communicate with each other [15].

2.3.2 PLC Regulations and Standards

2.3.2.1 Regulation Activities

The superimposing of high-frequency signals on PL causes electromagnetic radiation [17], which is more an issue in BB PLC that operates in high frequency bands. Therefore, regulations are imposed on PLC to limit the strength of the signals coupled into PL. For details of these regulations, refer to [18–20].

Some regulation bodies of NB PLC include the European Committee for Electrotechnical Standardization (CENELEC) bands, the U.S. FCC bands, the Japanese Association of Radio Industries and Businesses (ARIB) bands, and the Chinese Electric Power Research Institute (EPRI) bands. In particular, the CENELEC in Europe divides the NB PLC into four frequency bands, where band A (3–95 kHz) is reserved for energy providers, band B (95–125 kHz) is reserved for users, band C (125–140 kHz) is reserved for users as well while it is regulated to carrier-sense multiple-access with collision avoidance (CSMA/CA) mechanism, and band D (140–148.5 kHz) is reserved for users for alarm and security systems [19].

The regulation activities in BB PLC can be found in European Norm (EN) 50561-1 and Code of Federal Regulations, Title 47, Part 15 (47 CFR Part 15) by the U.S. FCC. In particular, EN 50561-1 applies to indoor PLC within the bandwidth of 1.6–30 MHz. The limits specified in these documents impose power spectral density (PSD) masks of around -55 dBm/Hz for transmission up to 30 MHz at an impedance of 100Ω, which is compatible with the PSD requirements in the standards ITU-T G. 9964 and IEEE 1901 [16].

2.3.2.2 Governing EMC Regulations

The EMC of the BPLC with the existing wireless devices is a crucial problem for the BPLC and gives rise to EMC broadband regulations [21]. The EMC regulations differ according to each country. For example, in Europe, the standard of Comité European de Normalisation Électrotechnique (CENELEC) organizes the emission measurement of each power line device. In addition, the CENELEC standard limits the PSD of the frequency spectrum below 30 MHz to -55 dBm/Hz and -85 dBm/Hz for the frequency band above 30 MHz. In United States, the limits of the emission for a given frequency is determined by the maximum radiated electric field that is measured at specific separation distance from the power line cable. In Japan, the standard of Comité International Spécial des Perturbations Radioélectriques (CISPR) referred to as CISPR22 limits the PSD for the band below 15 MHz to -71 dBm/Hz and -81 dBm/Hz for the band above 15 MHz.

2.3.2.3 Industrial and International Standards

IEEE 1901

In 2010, IEEE 1901 standard was released to organize the high-speed communication over the electric power lines. The standard spans the frequency band 1.8 MHz–50 MHz with 1974 subcarriers, where the subcarrier frequency spacing is 24.414 kHz. The OFDM symbol duration is taken as 40.96 μs with guard interval duration of 4.96 μs. The standard supports multiple spatial streams for both single- and three-phase power line cables.

HomeplugAV2

As mentioned in [21], the HomePlug Alliance, an industry-led organization, was formed in 2000 with the scope of promoting power line networking through the adoption of HomePlug specifications. In 2001, the HomePlug Alliance released the HomePlug 1.0.1 specification and followed up in 2005 with a second release: HomePlug AV. (The letters stand for audio and video.) Following its release, HomePlug AV rapidly became the most widespread adopted solution for in-home PLC. In December 2011, the Homeplug alliance issued the HomeplugAV2 standard [21] to provide the PHY and MAC layer for high-speed power line communications. HomeplugAV2 introduces two main enhancements compared to the IEEE 1901 standard [22]: (1) frequency band extension up to 86 MHz, and (2) MIMO capabilities with beamforming. This yields enhancing the achievable throughput in the HomeplugAV2 standard compared to the IEEE 1901 standard. The OFDM symbol uses 4,096 subcarriers in the band 1.8 MHz–100 MHz, where only 3,455 subcarriers are supported for communication in the band 1.8 MHz–86 MHz with subcarrier frequency spacing of 24.414 kHz. The OFDM symbol duration is 40.96 μs with GI duration of 4.96 μs. The HomeplugAV2 standard uses the same frame structure of IEEE 1901, which supports the carrier sense multiple access (CSMA) protocol. Each frame consists of four main fields: (1) preamble, (2) frame control (FC), (3) data or payload, and (4) selective acknowledge (SACK). The preamble field is responsible for packet synchronization functionality. The FC provides addressing and frame information. The FC and the preamble fields are called collectively the delimiter, which has the duration of 55.5 μs. For the data field, the maximum number of data bytes is 4095 bytes. The data field is followed by a gap period of time according to the CSMA protocol called the response interframe space (RIFS), where the maximum duration of the data field and the RIFS is 2501.12 μs. The SACK field is sent by the Rx to the Tx to request the retransmission of selected corrupted frames. The duration of the SACK is the same as the delimiter. The SACK field is followed by a gap period of time called the contention interframe space (CIFS). The CIFS and the RIFS periods are taken as 5 μs.

For the details on other standards and industry specifications of PLC, such as the Gigabit Home Networking (G.hn) standard and Powerline Related Intelligent Metering Evolution (PRIME), refer to [16,18–20].

2.4 TVWS Communications Interference to BPLC

For the cross talk between the two channels in the VHF band, the near-field coupling between broadband power line and high-frequency communication systems has been addressed in [23]. It depends on several factors like the cable properties, the coupling circuit, and the network asymmetric loads. Also, it has been proposed in [24] to extend the BPLC bandwidth to cover the whole VHF band of 30–300 MHz. However, BPLC in the VHF band suffers interference from surrounding wireless services such as TV and radio channels [25,26]. The current induced by incident electric field has been investigated thoroughly in many EMC applications [27–42].

From the measurement perspective, several measurement-based studies in [43–47] suggested enhancing the reception of the BPLC signal by adding an antenna to the receiver can compromise the radiated electric field. In other words, the authors in [43–47] suggested that by adding wireless antenna at the BPLC Rx, which can receive the radiated electric field during the BPLC transmission, and then adding the received wireless signal to the received BPLC signal, the signal-to-noise ratio of the received BPLC signal could be enhanced.

From a theoretical perspective, the electromagnetic field can interfere with the VHF BPLC by exciting the antenna-mode current [38], also known as the common-mode current [36]. In this case, the excited current has the same magnitude and phase, which implies that the sum of the TL currents is not equal to zero [38]. The excited antenna-mode current can convert to differential one by mode conversion mechanisms [48]. Those mechanisms depend mainly on the imbalance between TLs, which means that the impedance seen by each TL terminal to the ground is not the same [48]. This mode conversion is a significant source of interference to all the differential-mode communication signals such as VHF BPLC. In order to evaluate the power converted from the antenna mode to the differential mode at the TL terminals, the amount of the antenna mode power reflected or transmitted at the terminals have to be evaluated due to the mismatch between the terminal impedance and the antenna-mode characteristic impedance. This yields the importance of studying the antenna-mode characteristic impedance.

The main approach to describe the antenna-mode and differential-mode characteristic impedance is the TL theory, which assumes the existence of three conditions: (1) The cross section area of the wire is very small compared to the wavelength of the incident electric field, (2) The propagating field along the wire is either TEM or quasi-TEM, and (3) the sum of the currents propagating along the wire must be equal to zero. If any of those conditions is violated, then the TL theory approach becomes inadequate for modeling the characteristic impedance. In [37–42], an approximate TL theory solution was adopted to model the characteristic impedance of transmission lines.

When the first condition above is violated, several studies such as [39] and [40] have considered the dependence of the per unit length (p.u.l.) parameters on frequency. However, those studies did not consider the effect of TL length, which makes the solutions presented in [39] and [40] inadequate for the case of thin wire TLs excited by VHF electric field; that is, the case of electromagnetic interference which is induced across overhead lines or indoor power line cables. Also, the studies presented in [39] and [40] focused only on the differential-mode excitation and neglected the antenna-mode excitation, which is the main source of interference. In [41], an iterative method was proposed to solve the classical TL differential equations based on the perturbation theory. It was proved to have higher accuracy than the classical TL method in [42]. However, the iterative method could diverge at some frequencies. In [38], the authors provided a TL approximated solution for antenna-mode characteristic impedance. However, they did not take into consideration the effect of the frequency of operation or the angle of incidence of the electric field. In [49], a full-wave transmission line (FWTL) solution was proposed based on Maxwell's theory. The transmission line parameters were expressed using a parameter matrix. Iteration and perturbation methods were used to derive the exact solution for the differential-mode current case. The parameter matrix of the transmission line is updated for every iteration, which increases the complexity of the solution when being applied to the antenna-mode current case, since the integration formula of the parameter matrix gets more complicated for each iteration.

2.5 Enhancing the BPLC Capacity

Enhancing the BPLC capacity and satisfying the interference limit with wireless devices had been approached in the previous literature using three methods: (1) MIMO BPLC as in [21] and [50], (2) cognitive BPLC as in [51–55], and (3) hybrid wireless BPLC as in [56]. Although the MIMO BPLC can offer some enhancement to the system capacity, it still cannot offer high ergodic capacity in the VHF band due to the transmitter PSD mask forced by [22]. For cognitive BPLC, the BPLC Tx uses its coupling circuit to sense the spectrum before transmission to avoid interference with existing wireless services, and hence improve the overall ergodic capacity of the system. The coupling circuit is the circuit used to connect the BPLC transceiver to the power line network; it protects the transceiver from the mains voltage and filters the communication signal in the desired band. The main drawback in the cognitive BPLC is the weak reception capability of the BPLC coupling circuit to VHF wireless signals due to the near-field coupling loss between the BPLC and the VHF wireless channels as previously addressed in [23]. Besides, even for the MIMO cognitive BPLC, the coupling circuit itself is considered as a passive circuit device that can increase the coupling loss of the wireless signal. Also, weak coupling between the power line signals and the wireless VHF band can lead to a long sensing time that yields a capacity loss. In [43–47] preliminary studies were presented for sensing the wireless signal through an antenna, which can be added to the BPLC Rx. The measurement results in [44–46] prove that the wireless signal received by the antenna can be used in spectrum sensing for the BPLC communication. Also, in [47] further measurement results prove that the antenna can be used in the VHF band to exploit the radiation of the BPLC signal to enhance the reception diversity of the BPLC Rx.

As for hybrid wireless BPLC, WiFi was proposed in [56] to enhance the capacity of BPLC by building up a hybrid BPLC WiFi transceiver. The solution relies on the cooperation between the BPLC transceiver in the frequency band below 30 MHz with a transmitter PSD of −55 dBm/Hz [22] and the WiFi transceiver in the 2.45-GHz band with transmission power limit of 23 dBm [57], which is equivalent to −50 dBm/Hz for 20-MHz bandwidth. However, when the BPLC transceiver is used in the frequency band beyond 30 MHz, it suffers a lower transmission PSD limit of −85 dBm/Hz [22] while the WiFi transceiver in the 2.45-GHz band still has the same transmission power limit of 23 dBm. This limit difference results in allocation of most transmission power to the WiFi channel that suffers a higher path loss compared to the BPLC channel [58]. This leads to an overall capacity degradation compared to the BPLC WiFi capacity in the frequency band below 30 MHz.

2.6 BPLC Deployment into TVWS

2.6.1 Cooperation between BPLC and TVWS: Is It practical?

In Table 2.4, a summary is presented for the comparison between the TVWS and the BPLC in terms of the regulations, standards, and channel characteristics. The comparison shows the convergence between TVWS and BPLC in the following aspects. (1) The two technologies span common part of the VHF band, which enables cooperation between them to exploit that band. (2) The BPLC channel has the minimum coherence bandwidth of 65.5 kHz. Hence, utilizing the TVWS standard in the BPLC communication is applicable as the subcarrier frequency spacing is 53.5 kHz, which is less than the coherence bandwidth of the BPLC. (3) The TVWS standard introduces the minimum symbol duration of 13.8 μs. Hence, utilizing the TVWS as a standard in the BPLC channel is also applicable as the OFDM symbol duration is greater than the RMS delay of the BPLC channel. (4) The path loss introduced by the TVWS and BPLC channels lies approximately in the same range. Therefore, the MIMO techniques such as STBC or SM may be used to enhance the achievable throughput.

A cooperative BPLC TVWS system is illustrated in Figure 2.2. At the Tx side, each quadrature amplitude modulated (QAM) symbol is coded using space time block code (STBC). The output of the STBC is then split into two paths: (1) wireless TVWS, and (2) BPLC, where each path adopts OFDM to allocate specific frequency channels to each user. At the Rx, the received signals from the BPLC coupling circuit and the TVWS wireless antenna are combined using an STBC combiner. The combined symbols are passed through a maximum likelihood (ML) detector to allow estimation of the transmitted symbols. As illustrated in Figure 2.2, the system is within an office environment, where the sink can be connected to the main distribution box of the office and the other transceivers

Table 2.4
Comparison between TVWS and BPLC

	TVWS	BPLC
PSD (dBm/Hz)	−47.7 (nonadjacent channel) / −51.7 (adjacent channel)	−55 (<30 MHz) / −85 (>30 MHz)
Developed standards	IEEE 802.11af, IEEE 802.22 IEEE 802.15.4m, IEEE 802.19.1, IEEE 1900.7, ECMA-392	IEEE 1901, HomeplugAV1, HomeplugAV2
Frequency band (MHz)	54–806	1.8–100
Number of subcarriers	128 for each 6-MHz channel (ECMA-392)	3455 (HomeplugAV2)
Subcarrier frequency spacing (kHz)	53.5, 62.5, and 71.42	24.414
OFDM symbol duration (μs)	18.667, 16, and 13.8	40.96
GI duration (μs)	2.33, 2 and 1.725	4.96
Frame preamble and header duration (μs)	74.668, 64, and 55.2	55.5
Frame payload maximum number of bytes (byte)	4095	4095
Channel path loss at 100 MHz for coverage distance 15 m–50 m (dB)	51–81	38–81
Channel coherence bandwidth (kHz)	200	65.5–691.5
Channel RMS delay (ns)	100	413

can be connected to any outlet in the office. This creates point-to-multipoint communication case, which is further discussed in Chapter 7.

Cooperation between TVWS and BPLC can be achieved without adding up- or downconverter components (compared to other hybrid systems like the BPLC WiFi), since both technologies can access the VHF band. Hence, this cooperation has twofold benefits for TVWS and BPLC communication systems, which can be summarized as:

- For TVWS, BPLC supports the TVWS with MIMO capability in the VHF band, which improves significantly the achievable throughput and overcomes the aforementioned footprint design challenges in [6] and [7].
- BPLC supports the SU in the TVWS channels with a backup free channel in the high-frequency (HF) band (1.8 MHz–30 MHz). This facility can be utilized in the streaming services to minimize the total delay time.
- TVWS supports the BPLC communication in the VHF band with higher PSD under the umbrella of the FCC regulations. Hence, the power is

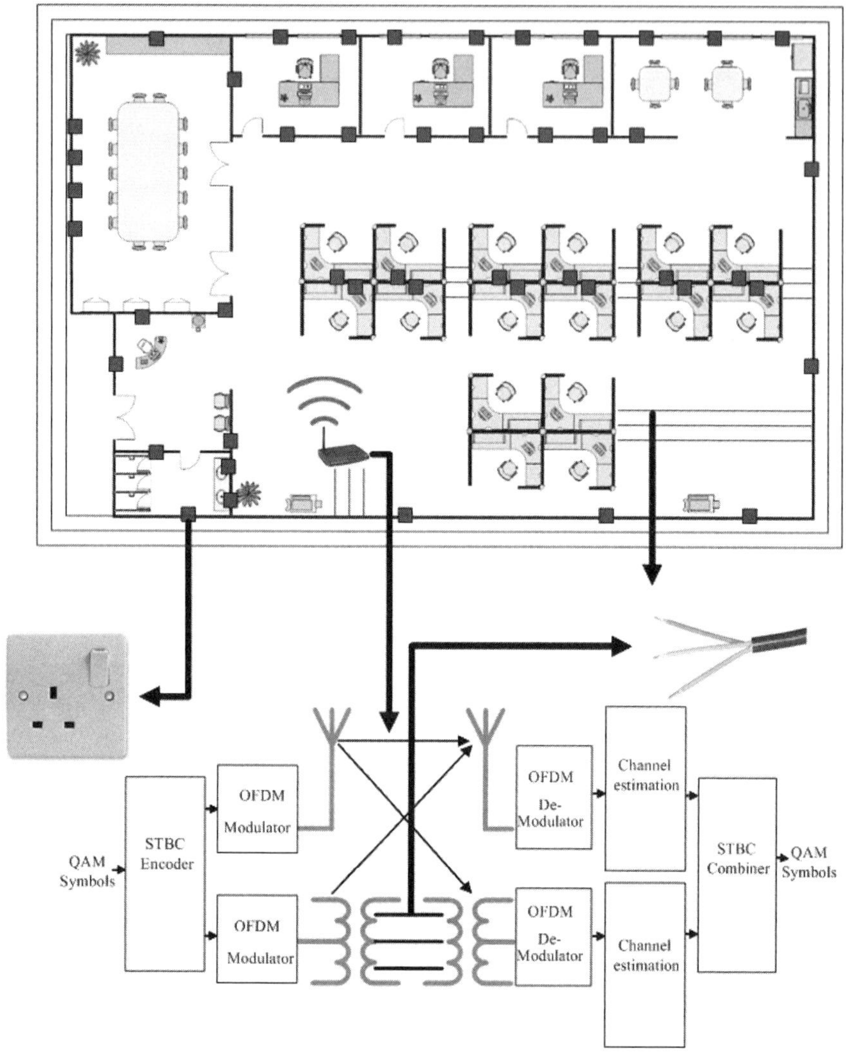

Figure 2.2 System model of cooperative BPLC TVWS system in an office environment.

efficiently distributed between the HF and VHF band, which yields a higher capacity.

- TVWS supports the BPLC with higher input power due to the increase in the PSD of the VHF band. Hence, using this power, the coverage area and the number of accommodated users can increase. It is worth mentioning that the maximum allowed input power for the BPLC is 282 mW, which is mostly concentrated in the HF band. Therefore, using higher power (which is feasible in a TVWS standard compliant system) can significantly improve the coverage area for point-to-multipoint communications in smart grid multimedia applications.

2.7 Summary

In this chapter, two indoor emerging technologies have been investigated. The regulations, standards, and the channel characteristics of each technology have been presented. A brief comparison between TVWS and BPLC was then made, which showed the feasibility of cooperation between these two technologies. TVWS has challenges in using MIMO techniques in the VHF band, while the BPLC suffers a high restriction on its PSD in the VHF band for EMC issues. Previous work done to enhance the BPLC capacity has been shown. Insufficiency has also been pointed out in the previous VHF BPLC capacity enhancement schemes. For the BPLC channel, a general path loss mapping is required to represent different BPLC topologies and deeper investigation is needed to model the interference of the wireless signal to the BPLC signal.

References

[1] Goldsmith, A., *Wireless Communications*, Cambridge, United Kingdom: Cambridge University Press, 2005.

[2] Maxwell, J. C., "A Dynamical Theory of the Electromagnetic Field," *Philosophical Transactions of the Royal Society of London*, Vol. 155, 1865, pp. 459–512.

[3] Federal Communication Commission, Second Report and Order and Memorandum Opinion and Order In the Matter of Unlicensed Operation in the TV Broadcast Bands, Additional Spectrum for Unlicensed Devices Below 900 MHz and in the 3 GHz Band, Document 08-260, November 2008.

[4] IEEE 802.11 af Draft 5.0, Amendment 5: TV White Spaces Operation. Retrieved April 2013.

[5] MAC and PHY for Operation in TV White Space, ECMA-392, December 2009.

[6] Tsakalaki, E., O. Alrabadi, E. De Carvalho, and G. F. Pedersen, "Antenna Design Considerations for MIMO TV White-Space Handsets, in *7th European Conference on Antennas and Propagation (EuCAP)*, April 2013, Gothenburg, Sweden, pp. 2597–2600.

[7] Nekovee, M., "Cognitive Radio Access to TV White Spaces: Spectrum Opportunities, Commercial Applications and Remaining Technology Challenges," in *IEEE Symposium on New Frontiers in Dynamic Spectrum*, April 2010, Singapore, pp. 1–10.

[8] Oh, J., S. Kim, and B. Jeong, "MIMO/Multi-Channel RF Transceiver for Cognitive Radio System," in *Proc. IEEE MTT-S International Microwave Workshop Series on Intelligent Radio for Future Personal Terminals (IMWS-IRFPT)*, Daejeon, South Korea, pp. 1–4, September 2011.

[9] Andrusenko, J., R. L. Miller, J. A. Abrahamson, et al., "VHF General Urban Path Loss Model for Short Range Ground-to-Ground Communications," *IEEE Transactions on Antennas and Propagation*, Vol. 56, No. 10, October 2008, pp. 3302–3310.

[10] Lin, Z., M, Ghosh, and A, Demir, "A Comparison of MAC Aggregation vs. PHY Bonding for WLANS in TV White Spaces," in *IEEE 24th International Symposium on Personal Indoor and Mobile Radio Communications (PIMRC)*, September 2013, London, pp. 1829–1834.

[11] Andrusenko, J., R. L. Miller, J. A. Abrahamson, N. M. M. Emanuelli, R. S. Pattay, and R. M. Shuford, "VHF General Urban Path Loss Model for Short Range Ground-to-Ground Communications," *IEEE Transactions on Antennas and Propagation*, Vol. 56, No. 10, October 2008, pp. 3302–3310.

[12] Brown, P. A., "Power Line Communications—Past, Present, and Future," in *Proc. IEEE International Symposium on Power-line Communications and Its Applications (ISPLC)*, Lancaster, United Kingdom, 1999, pp. 1–8.

[13] Schwartz, M., "Carrier-Wave Telephony over Power Lines: Early History [History of Communications]," *IEEE Communications Magazine*, Vol. 47, No. 1, 2009, pp. 14–18.

[14] Carcelle, X., *Power Line Communications in Practice*, Norwood, MA: Artech House, 2006.

[15] Galli, S., A. Scaglione, and Z. Wang, "For the Grid and through the Grid: The Role of Power Line Communications in the Smart Grid," *Proceedings of the IEEE*, Vol. 99, No. 6, 2011, pp. 998–1027.

[16] Cano, C., A. Pittolo, D. Malone, L. Lampe, A. M. Tonello, and A. G. Dabak, "State of the Art in Power Line Communications: From the Applications to the Medium," *IEEE Journal on Selected Areas in Communications*, Vol. 34, No. 7, 2016, pp. 1935–1952.

[17] See, K. Y., P. L. So, A. Kamarul, and E. Gunawan, "Radio-Frequency Common-Mode Noise Propagation Model for Power-Line Cable," *IEEE Transactions on Power Delivery*, Vol. 20, No. 4, 2005, pp. 2443–2449.

[18] Ferreira, H. C., L. Lampe, J. Newbury, and T. Swart (eds.), *Power Line Communications: Theory and Applications for Narrowband and Broadband Communications over Power Lines*, Chichester, United Kingdom: John Wiley & Sons, 2010.

[19] Lampe, L., Tonello, A. M., and T. G. Swart, *Power Line Communications: Principles, Standards and Applications from Multimedia to Smart Grid*, Chichester, United Kingdom: John Wiley & Sons, 2016.

[20] Galli, S., M. Koch, H. Latchman, et al., "Industrial and International Standards on PLC Base Networking Technologies," in *Power Line Communications: Theory and Applications for Narrowband and Broadband Communications over Power Lines*, H. C. Ferreira, L. Lampe, J. Newbury, and T. Swart (eds.), Chichester, United Kingdom: John Wiley & Sons, 2010, p. 377.

[21] Berger, L. T., A. Schwager, P. Pagani, and D. Schneider, *MIMO Power Line Communications: Narrow and Broadband Standards, EMC, and Advanced Processing*, Boca Raton, FL: CRC Press, 2014.

[22] IEEE Standards Association, IEEE Standard for Broadband over Power Line Networks: Medium Access Control and Physical Layer Specifications, IEEE Std 1901, 2010:1–1586, 2010.

[23] Karduri, M., M. D. Cox, and N. J. Champagne, "Near-Field Coupling between Broadband over Power Line (BPL) and High-Frequency Communication Systems," *IEEE Transactions on Power Delivery*, Vol. 21, No. 4, October 2006, pp. 1885–1891.

[24] Rehman, M. U., S. Wang, Y. Liu, S. Chen, X. Chen, and C. G. Parini, "Achieving High Data Rate in Multiband-OFDM UWB over Power-Line Communication System," *IEEE Transactions on Power Delivery*, Vol. 27, No. 3, July 2012, pp. 1172–1177.

[25] Khedimallah, S., B. Nekhoul, K. Kerroum, and K. El Khamlichi Drissi, "Analysis of Power Line Communications Electromagnetic Field in Electrical Networks Taking into Account the Power Transformers," in *Proc. IEEE International Symposium on Electromagnetic Compatibility*, September 2012, Rome, Italy, pp. 1–6.

[26] Wang, L. B., P. L. So, K. Y. See, M. Oswal, and T. S. Pang, "Investigation of Radiated Emissions in Power Line Communication Networks," in *Proc. IEEE International Power Engineering Conference (IPEC 2007)*, December 2007, Singapore, pp. 455–460.

[27] Tkachenko, S. V., R. Rambousky, and J. B. Nitsch, "Electromagnetic Field Coupling to a Thin Wire Located Symmetrically Inside a Rectangular Enclosure," *IEEE Transactions on Electromagnetic Compatibility*, Vol. 55, No. 2, April 2013, pp. 334–341.

[28] Rachidi, F., "A Review of Field-to-Transmission Line Coupling Models with Special Emphasis to Lightning-Induced Voltages on Overhead Lines," *IEEE Transactions on Electromagnetic Compatibility*, August 2012, Vol. 54, No. 4, pp. 898–911.

[29] Nucci, C. A., F. Rachidi, and M. Rubinstein, "Interaction of Lightning-Generated Electromagnetic Fields with Overhead and Underground Cables," *The Lightning Electromagnetics,* January 2012, pp. 687–718.

[30] Piantini A., and J. M. Janiszewski, "Lightning-Induced Voltages on Overhead Lines: Application of the Extended Rusck Model," *IEEE Transactions on Electromagnetic Compatibility*, Vol. 51, No. 3, August 2009, pp. 548–558.

[31] Tesche, F. M., and T. Karlsson, *EMC Analysis Methods and Computational Models*, New York: John Wiley & Sons, 1997.

[32] Ianoz, M., "Review of New Developments in the Modeling of Lightning Electromagnetic Effects on Overhead Lines and Buried Cables," *IEEE Transactions on Electromagnetic Compatibility,* Vol. 49, No. 2, May 2007, pp. 224–236.

[33] Bridges, G. E., "Transient plane Wave Coupling to Bare and Insulated Cables Buried in a Lossy Half-Space," *IEEE Transactions on Electromagnetic Compatibility*, Vol. 37, No. 1, February 1995, pp. 62–70.

[34] Petrache, E., F. Rachidi, M. Paolone, C. A. Nucci, V. A. Rakov, and M. A. Uman, "Lightning Induced Disturbances in Buried Cables—Part I: Theory," *IEEE Transactions on Electromagnetic Compatibility*, Vol. 47, No. 3, August 2005, pp. 498–508.

[35] Paolone, M., E. Petrache, F. Rachidi, et al., "Lightning Induced Disturbances in Buried Cables—Part II: Experiment and Model Validation," *IEEE Transactions on Electromagnetic Compatibility*, Vol. 47, No. 3, August 2005, pp. 509–520.

[36] Paul, C. R., "A Comparison of the Contributions of Common-Mode and Differential-Mode Currents in Radiated Emissions," *IEEE Transactions on Electromagnetic Compatibility*, Vol. 31, No. 2 May 1989, pp. 189–193.

[37] Poljak, D., A. Shoory, F. Rachidi, S. Antonijevic, and S. V. Tkachenko, "Time-Domain Generalized Telegrapher's Equations for the Electromagnetic Field Coupling to Finite Length Wires Above a Lossy Ground," *IEEE Transactions on Electromagnetic Compatibility*, Vol. 54, No. 1, February 2012, pp. 218–224.

[38] Vukicevic, A., F. Rachidi, M. Rubinstein, and S. V. Tkachenko, "On the Evaluation of Antenna-Mode Currents along Transmission Lines," *IEEE Transactions on Electromagnetic Compatibility*, Vol. 48, No. 4, November 2006, pp. 693–700.

[39] Maffucci, A., G. Miano, and F. Villone, "An Enhanced Transmission Line Model for Conducting Wires," *IEEE Transactions on Electromagnetic Compatibility*, Vol. 46, No. 4, November 2004, pp. 512–528.

[40] Chabane, S., P. Besnier, and M. Klingler, "Enhanced Transmission Line Theory: Frequency-Dependent Line Parameters and Their Insertion in a Classical Transmission Line Equation Solver," in *Proc. IEEE International Symposium on Electromagnetic Compatibility*, August 2013, Denver, Colorado, pp. 326–331.

[41] Rachidi, F., and S. Tkachenko, *Electromagnetic Field Interaction with Transmission Lines: From Classical Theory to HF Radiation Effects*, Volume 5, Southampton United Kingdom: WIT Press, 2008.

[42] Poljak, D., F. Rachidi, and S. V. Tkachenko, "Generalized Form of Telegrapher's Equations for the Electromagnetic Field Coupling to Finitelength Lines above a Lossy Ground," *IEEE Transactions on Electromagnetic Compatibility*, Vol. 49, No. 3, August 2007, pp. 689–697.

[43] Finamore, W. A., M. V. Ribeiro, and L.Lampe, "Advancing Power Line Communication: Cognitive, Cooperative, and MIMO Communication," in *Proc. Brazilian Telecommunications Symposium*, September 2012, Brasilia, Brazil, pp. 13–16.

[44] Degauque, P., P. Laly, V. Degardin, M. Lienard, and L, Diquelou, "Compromising Electromagnetic Field Radiated by In-House PLC Lines. In *Proc. IEEE Global Telecommunications Conference (GLOBECOM)*, IEEE, December 2010, Florida, pp. 1–5.

[45] Degardin, V., P. Laly, M. Lienard, and P. Degauque, "Compromising Radiated Emission from a Power Line Communication Cable," *Journal of Communications Software & Systems*, Vol. 7, No. 1, March 2011, pp. 16–21.

[46] Degauque, P., P. Laly, V. Degardin, and M. Lienard, "Power Line Communication and Compromising Radiated Emission," in *Proc. IEEE International Conference on Software, Telecommunications and Computer Networks (SoftCOM)*, IEEE, September 2010, Split-Bol, Croatia, pp. 88–91.

[47] Oliveira, T. R., C. A. G. Marques, M. S. Pereira, S. L. Netto, and M. V. Ribeiro, "The Characterization of Hybrid PLC-Wireless Channels: A Preliminary Analysis," in *Proc. 17th IEEE International Symposium on Power Line Communications and Its Applications (ISPLC)*, IEEE, March 2013, Johannesburg, South Africa. pp. 98–102.

[48] Kaiser, K. L., *Electromagnetic Compatibility Handbook*, Boca Raton, FL: CRC Press, 2005.

[49] Nitsch, J., and S. Tkachenko, "High-Frequency Multiconductor Transmission-Line Theory." *Foundations of Physics*, Vol. 40, No. 9–10, 2010, pp. 1231–1252.

[50] Weling, N., "SNR Based Detection of Broadcast Radio Stations on Powerlines as Mitigation Method Toward a Cognitive PLC Solution," in *Proc. 16th IEEE International Symposium on Power Line Communications and Its Applications (ISPLC)*, IEEE, March 2012, Beijing, China, pp. 52–59.

[51] Vuohtoniemi, R., J.-P. Makela, J. Vartiainen, and J. Iinatti, "Detection of Broadcast Signals in Cognitive Radio Based PLC Using the FCME Algorithm," in *Proc. IEEE International

Symposium on Power Line Communications and its Applications (ISPLC), March 2014, Glasgow Scotland, pp. 70–74.

[52] Praho, B., M. Tlich, P. Pagani, A. Zeddam, and F. Nouvel, "Cognitive Detection Method of Radio Frequencies on Power Line Networks," in *Proc. IEEE International Symposium on Power Line Communications and Its Applications (ISPLC)*, March 2010, Rio de Janeiro, Brazil, pp. 225–230.

[53] Ali, K. M., G. G. Messier, and S. W. Lai, "DSL and PLC Co-Existence: An Interference Cancellation Approach," *IEEE Transactions on Communications*, Vol. 62, No. 9, September 2014, pp. 3336–3350.

[54] Oh, S. W., Y. L. Chiu, K. N. Ng, et al., "Cognitive Power Line Communication System for Multiple Channel Access," in *Proc. IEEE International Symposium on Power Line Communications and Its Applications*, March 2009, Dresden Germany, pp. 47–52.

[55] Maes, J., M. Timmers, and M. Guenach, "Spectral Compatibility of In-Home and Access Technologies," in *Proc. IEEE International Symposium on Power Line Communications and Its Applications (ISPLC)*, April 2011, Udine Italy, pp. 7–11.

[56] Lai, S. W., N. Shabehpour, G. G. Messier, and L. Lampe, "Performance of Wireless/Power Line Media Diversity in the Office Environment," in *Proc. IEEE Global Communications Conference (GLOBECOM)*, December 2014, Texas, pp. 2972–2976.

[57] O'Hara, B., and A. Petrick. *IEEE 802.11 Handbook: A Designer's Companion.* New York: IEEE Standards Association, 2005.

[58] Ali, A. H., M. R. A. Razak, M. Hidayab, , et al., "Investigation of Indoor WIFI Radio Signal Propagation," in *Proc. IEEE Symposium on Industrial Electronics & Applications (ISIEA) 2010*, October 2010, Penang, Malaysia, pp. 117–119.

3

Channel Models of Wireless and Power Line Communications

3.1 Wireless Communication Channels

The wireless channel serves as the physical transmission medium between the signal transmitter and receiver. Its characteristics have a dominant impact on the performance of the wireless communication systems, such as the maximum achievable channel capacity and the outage probability. Therefore, it is of paramount importance to understand the characteristics of wireless channel to enable reliable high-speed communication.

3.1.1 Large-Scale Propagation

Large-scale propagation involves variations of the signal strength at the receiver over distance caused by path loss and shadowing. Typically, these variations occur over relatively large distances (100–1,000m in the case of path loss and 10–100m in the case of shadowing in outdoor scenario and less in indoor scenario), and hence the name large-scale propagation [1].

3.1.1.1 Path Loss

Path loss is the reduction of signal strength owing to the separation of large distance between the transmitter and the receiver. In wireless communications, the transmitter radiates the energy of a signal, which then propagates through a certain medium in the form of electromagnetic waves. If part of the energy can be received at the receiver side, the information carried within the signal can possibly be recovered correctly, therefore establishing a wireless communication link.

Assuming the radiation pattern of the antenna is isotropic (i.e., the antenna radiates signal equally in all directions) and the electromagnetic wave propagates in the vacuum environment, then the electromagnetic wave propagates in the pattern of a sphere. The energy conservation law [2] indicates that the total energy of the radiated signal remains constant through any sphere centered at the transmitter. Therefore, in the case of isotropic antenna and free-space propagation, the energy at the unit area on the spherical surface centered at the transmitter decreases with the increase of the distance that the electromagnetic wave travels (the area of the spherical surface increases). Denoting the radius of the spherical surface centered at the transmitter as d, and the transmit antenna is isotropic, then the power density, P_{eff}, on the surface can be given as [2]

$$P_{eff} = \frac{P_t}{4\pi d^2} \tag{3.1}$$

where P_t is the transmit power. In the case that the transmit antenna is not isotropic, the transmit antenna gain G_t in the direction of the receive antenna should be taken into account in the expression of P_{eff} as [2]

$$P_{eff} = \frac{G_t P_t}{4\pi d^2} \tag{3.2}$$

The effective aperture (shown in Figure 3.1) is a measure of the effectiveness of an antenna at receiving the power of electromagnetic waves. Mathematically, it can be expressed as [2]

$$A_{eff} = \frac{\lambda^2}{4\pi} G_r \tag{3.3}$$

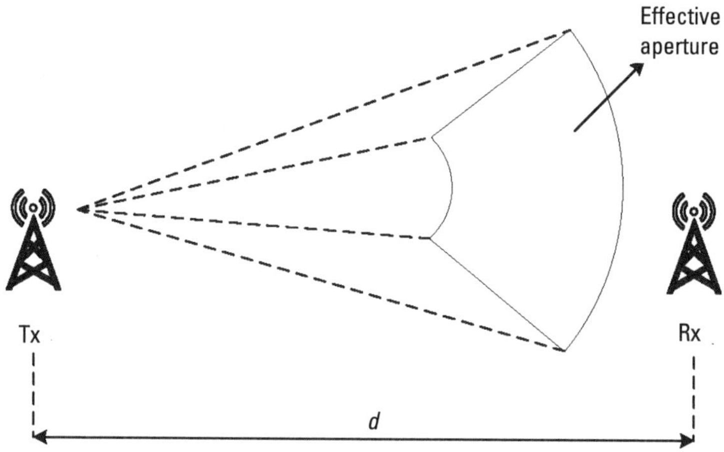

Figure 3.1 Propagation of the electromagnetic wave (d denotes the distance between Tx and Rx).

where λ is the wavelength of the operation frequency, and G_r is the receive antenna gain. According to the Friis' law [2], the received power P_r can be determined as

$$P_r = P_{eff} \times A_{eff} = \frac{P_t G_t G_r \lambda^2}{(4\pi)^2 d^2} \tag{3.4}$$

Therefore, the free-space linear path loss (defined as the ratio of the transmit power to the receive power [2]) can be calculated as

$$PL = \frac{P_t}{P_r} = \frac{(4\pi)^2 d^2}{G_t G_r \lambda^2} \tag{3.5}$$

More generally, the linear path loss can be expressed as

$$PL = \frac{P_t}{P_r} = \frac{(4\pi)^2 d^n}{G_t G_r \lambda^2} \tag{3.6}$$

where n is the path loss exponent. Typically, the value of the path loss exponent can vary between 1.6 to 6 based on different propagation environments (shown in Table 3.1).

Practically, the path loss often appears in the log form (dB value of the linear path loss). Therefore, (3.6) can be expressed in dB as

$$PL(dB) = 10\log_{10}\frac{P_t}{P_r} = 10\log_{10}\left[\frac{(4\pi)^2 d^n}{G_t G_r \lambda^2}\right] \tag{3.7}$$

which can also be rewritten as

$$PL(dB) = C + 10n\log_{10}d \tag{3.8}$$

where $C = 10\log_{10}\left[\frac{(4\pi)^2}{G_t G_r \lambda^2}\right]$. In addition, a simplified path loss model as a function of distance is often used in designing a wireless system, the expression

Table 3.1
Path Loss Exponent Values in Different Environments

Environment	Path Loss Exponent n
In building line-of-sight	1.6 to 1.8
Free space	2
Obstructed in factories	2 to 3
Urban area cellular	2.7 to 3.5
Shadowed urban cellular	3 to 5
Obstructed in building	4 to 6

can be given as [1]

$$P_r = P_t K \left[\frac{d_0}{d} \right]^n \tag{3.9}$$

which can be written in the log form as

$$P_r(dB) = P_t(dB) + 10 \log_{10} K + 10n\log_{10} \left[\frac{d_0}{d} \right] \tag{3.10}$$

where K is a constant that is related to the antenna characteristics and the average channel attenuation, n denotes the path loss exponent, d_0 represents the reference distance for the antenna far field, and d is the distance of interest from the transmitter. Since distance affects the path loss greatly, the path loss plays a major role in designing and analyzing a wireless communication system.

3.1.1.2 Shadowing

A signal may encounter obstacles (such as buildings, trees, and walls) in its path of propagation. These obstacles cause effects such as reflection,[1] diffraction,[2] and scattering[3] of the signal and thus introduce random variations about the path loss. A widely adopted model for such random variations is the lognormal shadowing model, which has been empirically proven to be accurate in both indoor and outdoor radio propagation scenarios [1].

The effect of path loss and shadowing can be captured by a combined model that superimposes these two components. In the combined model, the average path loss is represented by the path loss model (equation for path loss is shown in (3.10)) and random variations due to shadowing are then added into the model. Specifically, the ratio of the received power to the transmitted power in the log form in the combined model can be expressed as [1]

$$\frac{P_r}{P_t}(dB) = 10 \log_{10} K + 10n \log_{10} \left[\frac{d_0}{d} \right] + X_\sigma \tag{3.11}$$

where X_σ is a Gauss-distributed random variable with a zero mean and variance of σ^2, which is caused by the effect of shadowing.

3.1.2 Small-Scale Propagation

Small-scale propagation effect or multipath fading is the rapid fluctuations of the amplitudes, phases, or multipath delays of a wireless signal over a short period of

[1] Reflection occurs when radio waves encounter a surface with dimensions that are relatively large compared to the wavelength of the signal, such as the surface of buildings.

[2] Diffraction occurs when radio waves propagate around the sharp edge of an object that is impenetrable with dimensions larger than the wavelength of the signal, such as street corners.

[3] Scattering occurs when radio waves encounter an object with dimensions in the order of the wavelength of the signal or less, such as raindrops.

time or distance (on the order of the signal wavelength, and hence the name small-scale fading), in which case the large-scale path loss effects could be ignored [1]. Small-scale fading is typically due to multipath propagation, which is often the case in urban environments. In such environments, the presence of buildings, trees, vehicles, pedestrians, and so forth, around the transmitter and the receiver could cause reflection, diffraction, and scattering of the transmitted signal. Consequently, the transmitted signal arrives at the receiver through different paths, and each of the paths generate a unique wave of the transmitted signal with random variations of the amplitude, phase, and delay. Finally, all the waves undergoing different paths are combined at the receiver and cause distortion and fading to the transmitted signal.[4] In the following, factors and parameters relating to small-scale fading will be introduced.

3.1.2.1 Root Mean Square Delay Spread

In wireless communications, the transmitted signal undergoes different paths and therefore arrives at the receiver at different times of arrival. The delay spread is a measure of the difference between the time of arrival of the earliest multipath component and the time of arrival of the latest multipath component. The delay spread can be characterized by the power delay profile, which is the distribution of received signal power as a function of the propagation delays, and can be obtained empirically [1]. Denoting g_i and τ_i as the path gain and delay for the ith path of a transmitted signal, respectively, then the mean excess delay $\bar{\tau}$, which is the first moment of the power delay profile, can be defined as [3]

$$\bar{\tau} = \frac{\sum_i g_i^2 \tau_i}{\sum_i g_i^2} \tag{3.12}$$

The RMS delay spread is the square root of the second central moment of the power delay profile [3] and can be determined as

$$\sigma = \sqrt{\bar{\tau^2} - (\bar{\tau})^2} \tag{3.13}$$

where

$$\bar{\tau^2} = \frac{\sum_i g_i^2 \tau_i^2}{\sum_i g_i^2} \tag{3.14}$$

These delays are measured relative to the first detectable signal arriving at the receiver at $\tau_0 = 0$. If there is only one path, the RMS delay spread would be zero. The RMS delay spread is a representation of the effect of multipath; that is, a higher RMS delay spread value indicates a larger effect of the multipath.

---

[4]The shadowing effect reflects the fluctuation of a wireless signal over relatively large distances (10–100 m in an outdoor scenario and less in an indoor scenario), while a small-scale propagation effect reflects the fluctuation of a wireless signal on the order of the signal wavelength.

3.1.2.2 Coherence Bandwidth

The coherence bandwidth is a statistical measure of the range of frequencies over which the channel can be considered flat; that is, a channel passing all spectral components with approximately equal gain and linear phase. As well, or coherence bandwidth is the range of frequencies where two frequency components have a strong potential correlation for amplitude [3]. The coherence bandwidth can be derived from the RMS delay spread, while the exact relationship between the coherence bandwidth and the RMS delay spread depends on the definition. If the coherence bandwidth is defined as the bandwidth over which the frequency correlation function is above 0.9, then the coherence bandwidth is approximately [3]

$$B_c \approx \frac{1}{50\sigma} \tag{3.15}$$

where σ is the RMS delay spread. While in the case that the coherence bandwidth is defined as the bandwidth over which the frequency correlation function is above 0.5, then the coherence bandwidth is approximately [3]

$$B_c \approx \frac{1}{5\sigma} \tag{3.16}$$

Note that the coherence bandwidth and the RMS delay spread are inversely related. An increase of the RMS delay spread value will result in a decrease of the coherence bandwidth.

3.1.2.3 Doppler Spread

Doppler spread, B_d, is a measure of the broadening in spectrum due to the time-varying nature of the mobile radio channel. For example, in the case of a mobile radio channel, when a pure sinusoidal wave with frequency f_c is transmitted, the received signal spectrum (or the Doppler spectrum[5]) will have components ranging from $f_c - f_d$ to $f_c + f_d$, where f_d denotes the Doppler shift.[6] Specifically, the Doppler shift is a function of the relative velocity of the mobile v, the wavelength of the carrier frequency λ, and the angle between the direction of the mobile user's motion and the direction of the received signal wave θ. Mathematically, the Doppler shift can be expressed as [3]

$$f_d = \frac{v}{\lambda}\cos\theta \tag{3.17}$$

In wireless communications, a transmitted signal may undergo different paths through its way to the receiver, and each path will induce different Doppler shifts. The maximum Doppler shift is achieved when the direction of the mobile

[5] Doppler spread can be regarded as the bandwidth of the Doppler spectrum.

[6] Doppler shift is caused by the movement of the user (or surrounding physical presences in the environment). The user's velocity will induce a shift in the frequency of the transmitted signal along each signal path.

user's motion is in accordance with the direction of the received signal wave, with $\theta = 0$ and $f_{max} = \frac{v}{\lambda}$. The difference in Doppler shifts between different signal components is referred to as the Doppler spread.

3.1.2.4 Coherence Time

The coherence time, T_c, is the time duration over which the channel impulse response remains invariant, or the time duration over which two received signals have a strong potential correlation in amplitude [3]. Generally, the coherence time can be employed to characterize the time-varying nature of the frequency dispersiveness of the channel in the time domain. The coherence time and Doppler spread are inversely proportional to each other, as given by [3]

$$T_c \approx \frac{1}{f_{max}} \tag{3.18}$$

where f_{max} is the maximum Doppler shift. In the case that the coherence time is defined as the time over which the time correlation function is above 0.5, and then the mathematical expression for coherence time becomes [3]

$$T_c \approx \frac{9}{16\pi f_{max}} \tag{3.19}$$

The definition of coherence time suggests that two signals arriving with a time separation larger than T_c will be affected differently by the channel.

3.1.2.5 Types of Small-Scale Fading

With the factors and parameters relating to small-scale fading introduced (i.e., RMS delay spread, coherence bandwidth, Doppler spread, and coherence time), different types of small-scale fading can be categorized based on the thresholds of the relating factors and parameters together with the characteristics of the transmitted signal (e.g., signal bandwidth and symbol period). Generally, four types of fading effects can be categorized: flat fading, frequency selective fading, fast fading, and slow fading [1]. These fading types and the corresponding conditions are shown in Table 3.2. In the following, these types of fading will be explained in detail.

- *Flat fading*: If the symbol period of a transmitted signal, T_s, is larger than the RMS delay spread of a wireless channel, σ, or the bandwidth

Table 3.2
Small-Scale Fading Types

Flat fading	Signal bandwidth $<$ Coherence bandwidth
Frequency selective fading	Signal bandwidth $>$ Coherence bandwidth
Fast fading	Coherence time $<$ symbol duration
Slow fading	Coherence time $\gg$ symbol duration

of the transmitted signal, B_s, is smaller than the channel coherence bandwidth, B_c, then the channel is referred to as a flat fading channel [1]. In a flat fading channel, the channel exhibits approximately the same channel gain and linear phase throughout the transmission bandwidth. The channel gains of a flat fading channel is time-varying according to certain distributions such as Rician fading, Rayleigh fading, and Nakagami fading [3].

- *Frequency selective fading*: Contrary to the flat fading channel, if the symbol period of a transmitted signal, T_s, is smaller than the RMS delay spread of a wireless channel, σ, or the bandwidth of the transmitted signal, B_s, is larger than the channel coherence bandwidth, B_c, then the channel is referred to as a frequency selective fading channel [1]. In a frequency selective fading channel, the previous transmitted symbols can cause interference to the current transmitted symbols, as the waves with long delays of the previous transmitted symbols and the waves with short delays of the current transmitted symbols can arrive at the receiver concurrently. This interference is known as intersymbol interference (ISI). In the frequency domain, the ISI is presented by a formation that frequency components of the received signal spectrum have different amplitudes.
- *Fast fading*: If the symbol period of a transmitted signal, T_s, is larger than the channel coherence time, T_c, or the bandwidth of the transmitted signal, B_s, is smaller than the Doppler spread, B_d, then the channel is referred to as a fast fading channel [1]. In a fast fading channel, the channel changes so rapidly that the signal undergoes different channels within the time duration of one symbol period.
- *Slow fading*: If the symbol period of a transmitted signal, T_s, is much smaller than the channel coherence time, T_c, or the bandwidth of the transmitted signal, B_s, is much larger than the Doppler spread, B_d, then the channel is referred to as a slow fading channel [1]. In a slow fading channel, the channel remains static over one or several continuous symbol periods.

Flat and frequency selective fading determine the frequency diversity of the channel while fast and slow fading determine the time diversity of the channel. Based on the different fading effects, four small-scale fading channels can be categorized: flat fast fading channel, flat slow fading channel, frequency selective fast fading channel, and frequency-selective slow fading channel. These four channels are illustrated in Figures 3.2 and 3.3 as classified based on the symbol period, T_s, with the RMS delay spread, σ, and the channel coherence time, T_c, as thresholds, and based on the symbol bandwidth, B_s, with the

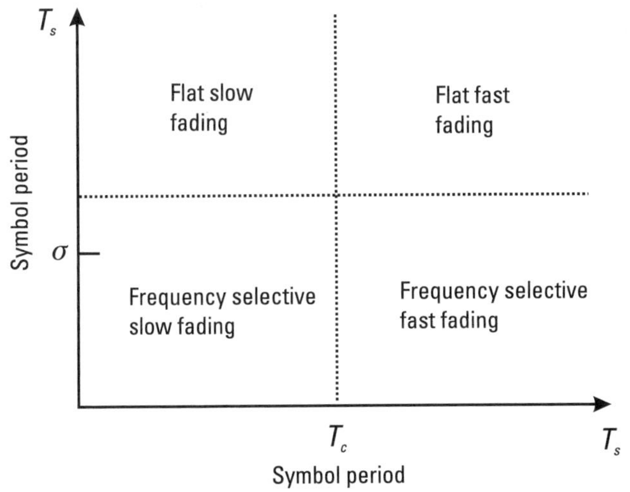

Figure 3.2 Fading types classified by symbol period.

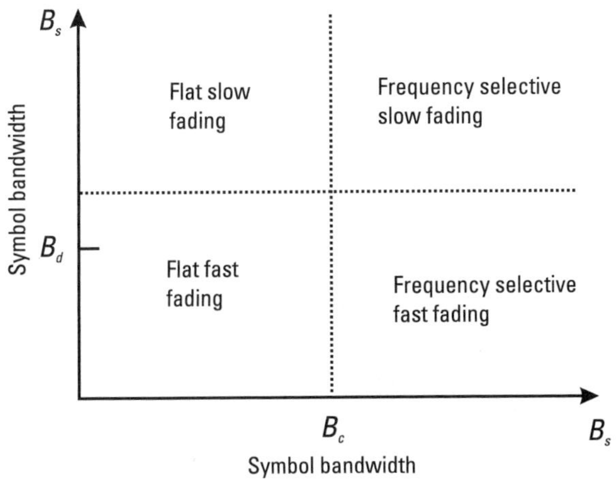

Figure 3.3 Fading types classified by symbol bandwidth.

Doppler spread, B_d, and the channel coherence bandwidth, B_c, as thresholds, respectively.

3.1.3 Additive White Gaussian Noise Channel

The wireless communication channel has various channel impairments that could cause errors. Noise, which is an inevitable component in the design and analysis of the performance of a communication system, refers to the undesired signals that would affect the fidelity of the desired signal. In wireless communication systems, there are various types of noise, such as thermal noise and impulse noise. In the case of direct LoS path between the transmitter and the receiver, the additive white Gaussian noise (AWGN) channel can provide a reasonably good model. In the AWGN channel, the noise is always assumed to be white, which indicates that all frequency components in the noise exhibit the same power and the transmitted signal is added with a noise component that has a Gaussian distribution with zero-mean. Mathematically, the AWGN channel model can be expressed as [1]

$$r(t) = s(t) + n(t) \tag{3.20}$$

where $s(t)$ is the transmitted signal, $n(t)$ is the zero-mean white Gaussian noise process with the power spectral density of N_0, $r(t)$ is the received signal, and t denotes the time.

Assuming the AWGN channel has a bandwidth of B and a transmit power of P, then the capacity of this channel, C, can be provided by Shannon's formula [4]:

$$C = B \log_2(1 + \frac{P}{N_0 B}) \tag{3.21}$$

where N_0 is the power spectral density of the noise.

3.2 Power Line Communication Channels

3.2.1 PLC Channel Modeling

Initially designed for the purpose of electricity distribution, the PL channel is a very harsh medium for data communication. Generally, it suffers from frequency selective fading that is caused by signal reflections due to impedance mismatch between different types of loads and PL cables [5]. It is also prone to topology changes caused by factors such as the plug in/out of appliances, which makes the PL channel time-varying. Various sources of noise in PLC further degrade the performance of the PL channel [6]. Therefore, the modeling of PLC channel to explore PLC channel characteristics, which facilitates the development of advanced technology to combat these PLC channel impairments, has become an inevitable necessity.

Research efforts on PLC channel modeling in recent years have gone in two directions, the bottom-up approach (or deterministic approach) [7–10] and the top-down approach (or statistical approach) [11–13]. The bottom-up approach exploits the transmission line (TL) theory to compute the channel transfer function (CTF) and henceforth obtains a frequency response of the PL channel. The CTF can be obtained deterministically based on the network topology. While the bottom-up approach captures physical reality, it has a rather high computational complexity. The top-down approach, however, treats the PL channel as a black box and the channel response is obtained by fitting certain statistical data from experiment into the channel parametric function [11]. This

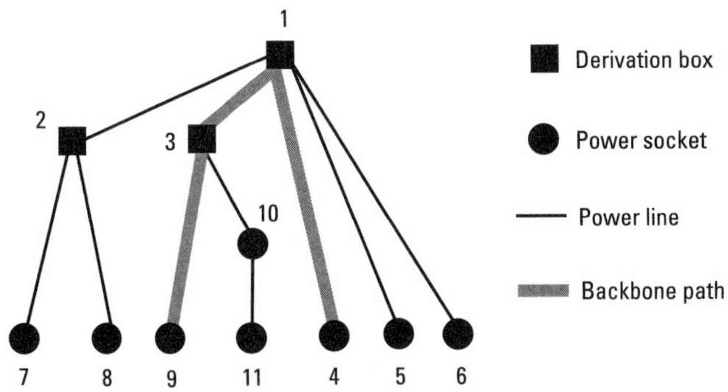

Figure 3.4 Backbone path identification.

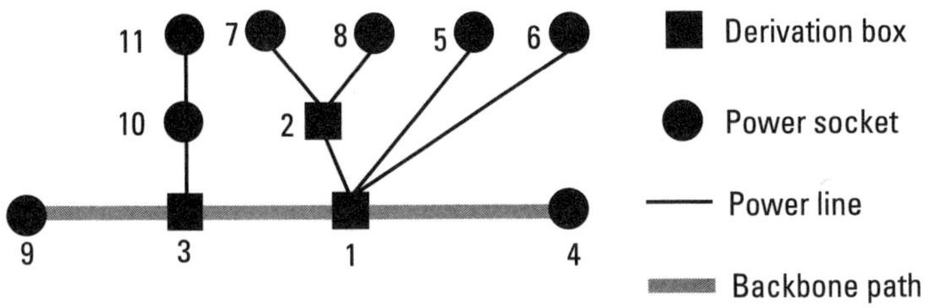

Figure 3.5 Topology remapping along the backbone path.

approach enables a fast channel simulation at the cost of lossy link to the physical reality.

3.2.1.1 Bottom-Up Approach

The bottom-up approach is developed based on the TL theory [14], in which complete information of the network topology is required. In order to calculate the CTF between any pairs of power sockets in the PLC network, backbone path (the shortest path between the pair of power sockets) identification and topology remapping is necessary. Figure 3.4 shows the identification of the backbone path for a randomly generated topology (details on random indoor PLC topology generator can be found in [7]) to obtain the CTF between node 9 and node 4. Figure 3.5 depicts the remapped PLC topology along the backbone path.

To simplify the calculation of CTF, the impedances that are not on the backbone path are carried back on the backbone path exploiting TL theory (i.e., node 2, 5–8, and 10, 11 in Figure 3.5). To start with, it is assumed a transverse electromagnetic (TEM) or quasi TEM propagation mode[7] [7]. As shown in

[7]In the transverse electric (TE) mode, the electric vector (E) is always perpendicular to the direction of propagation, while there is only a magnetic field along the direction of propagation. In the transverse magnetic (TM) mode, the magnetic vector (H) is always perpendicular to the direction of propagation, while there is only an electric field along the direction of propagation. In

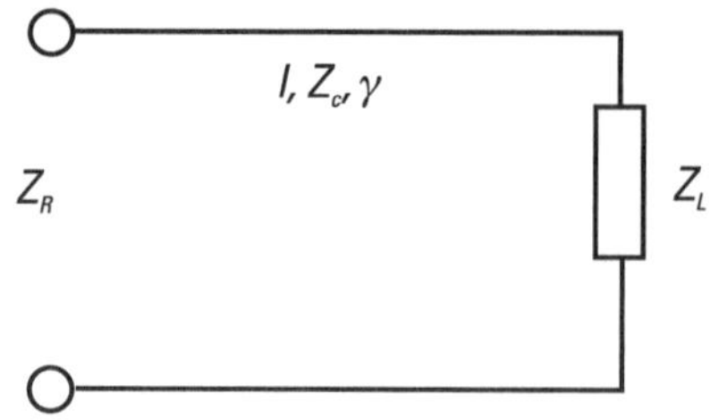

Figure 3.6 Illustration of a section of transmission line terminated with load.

Figure 3.6, for a TL of the length l, characteristic impedance Z_C and propagation constant γ, with an impedance load Z_L, the equivalent impedance Z_R from the line input can be calculated exploiting TL theory [7]:

$$Z_R = Z_C \frac{Z_L + Z_C \tanh(\gamma l)}{Z_C + Z_L \tanh(\gamma l)} = Z_C \frac{1 + \rho_L e^{-2\gamma l}}{1 - \rho_L e^{-2\gamma l}} \tag{3.22}$$

where $\rho_L = (Z_L - Z_C)/(Z_L + Z_C)$. The above equation is used recursively for impedance carry-back.

To calculate the CTF, considering a backbone portion that consists an output node (toward the receiver terminal) and an input node (toward the transmitter terminal), if the characteristic impedance of a transmission line segment is denoted as Z_{C_b}, the propagation coefficient is denoted as γ_b, the length of the associated backbone line unit is l_b, and the load at the output node is Z_{L_b}, then the load reflection coefficient for the backbone portion is calculated as [7]

$$\rho_{L_b} = \frac{Z_{L_b} - Z_{C_b}}{Z_{L_b} + Z_{C_b}} \tag{3.23}$$

Then, based on the voltage ratio approach (VRA) [7], the CTF for a backbone portion can be obtained by

$$H_b = \frac{V_b}{V_{b-1}} = \frac{1 + \rho_{L_b}}{e^{\gamma_b l_b} + \rho_{L_b} e^{-\gamma_b l_b}} \tag{3.24}$$

where V_{b-1} and V_b are the voltage at the input node and output node, respectively. If the backbone is divided into $N + 1$ units, and denoting f as the operating frequency, the total CTF can be computed as [7]

$$H(f) = \prod_{b=1}^{N+1} H_b(f) \tag{3.25}$$

the TEM mode, both the electric vector (E) and the magnetic vector (H) are perpendicular to the direction of propagation.

where the characteristic impedance of the transmission line segment, Z_{C_b}, and the propagation coefficient, γ_b, are frequency-dependent.

Other than VRA, ABCD matrix method [15] can also be used to calculate the CTF.

3.2.1.2 Top-Down Approach

The bottom-up approach is able to obtain deterministic CTF with the requirement of global information of all elements in the PLC network (e.g., PL cable types, PL cable lengths, load impedances). In particular, when a large scale of PLC network is considered, it would incur tremendous computational complexity, not to mention that the required global information is usually difficult, if not impossible, to obtain [11]. In light of the drawbacks of the bottom-up approach, a top-down approach is advocated by many studies [11–13] due to much reduced complexity. The top-down approach treats the PL channel as a black box [11], of which the general statistics of CTF is studied based on extensive experimental measurements. By analyzing the measurement results, general channel parametric function is derived to obtain CTF by data fitting, which requires very few relevant parameters (or statistical data) of the PLC network.

A widely adopted top-down approach to model the CTF is reported in [11], in which the effects of multipath propagation, cable length, multipath delay, and frequency selective attenuation are jointly considered. In this model [11], the frequency response of CTF is given by

$$H(f) = \sum_{i=1}^{N} g_i \cdot e^{-(a_0+a_1 \cdot f^k)\cdot d_i} \cdot e^{-j2\pi f(d_i/v_p)} \tag{3.26}$$

where N is the number of paths, and g_i is a weighting factor that is related to the reflection and transmission factors along a propagation path. a_0 and a_1 are attenuation parameters, while k is the exponent of the attenuation factor with typical values between 0.5 and 1. These three parameters are typically derived from measurement statistics of the CTF. d_i and f indicate the length of path i and the frequency, respectively. The second term, $e^{-(a_0+a_1 \cdot f^k)\cdot d_i}$, as a whole accounts for the attenuation on the ith path, which becomes more severe with the increase of d_i and f. v_p represents the propagation speed and $\frac{d_i}{v_p}$ indicates the delay τ_i on the ith path. The last portion of this model, $\sum_{i=1}^{N} e^{-j2\pi f(d_i/v_p)}$, accounts for the multipath delay.

In the PL signal propagation, it does not only occur along a direct LoS path between the transmitter and the receiver, but additional paths (signal reflections) should also be taken into account, which results in a multipath signal propagation scenario. For example, in the network topology shown in Figure 3.7, considering

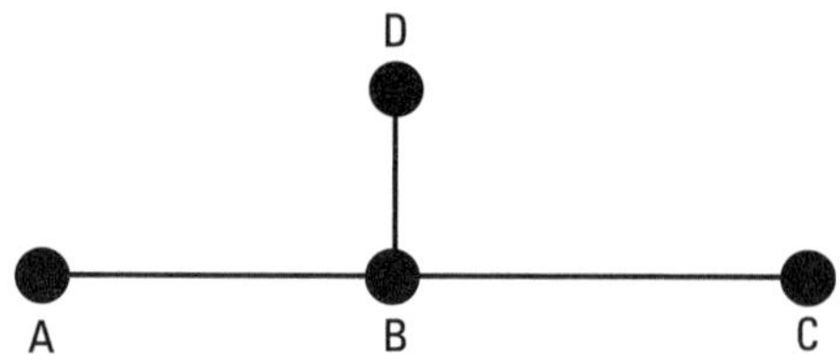

Figure 3.7 PLC multipath signal propagation.

point A as the transmitter and C as the receiver, then the signal propagation paths due to multiple reflections could be $A \to B \to C$, $A \to B \to D \to B \to C$, $A \to B \to D \to B \to D \to B \to C$, and so on. The signal propagation undergoing different paths would arrive at the receiver at different times, and hence cause multipath delay.

The top-down model allows fast CTF generation with statistical data fitting. However, extensive experimental measurements are required to determine the parameters (e.g., a_0, a_1, and k). In [16], the Open PLC European Research Alliance (OPERA) classed nine reference channels as representatives of real PL channels for the LV/MV network. Later in [17], nine sets of parameters are reported corresponding to the nine reference channels in [16], which allows the fast statistical generation of nine class CTFs.

3.3 Summary

In this chapter, the channel models for wireless and power line communications were reviewed. Although random topology change, the time-varying and frequency-selective fading nature, and the complex noise scenarios of PLC make PL a harsh medium for data communication, some advanced PLC technologies such as PLC topology inference [18–21], OFDM PLC [22–24], and PLC noise detection and mitigation [25–28] have been developed to combat the channel impairments of PLC.

References

[1] Goldsmith, A., *Wireless Communications*, Cambridge, United Kingdom: Cambridge University Press, 2005.

[2] Molisch, A. F, *Wireless Communications*, Chichester, United Kingdom: John Wiley & Sons, 2012.

[3] Rappaport, T. S., *Wireless Communications: Principles and Practice*, Prentice Hall PTR, 2002.

[4] Shannon, C. E., "A Mathematical Theory of Communication," *Bell System Technical Journal*, Vol. 27, 1948, pp. 623–656.

[5] Ferreira, H. C., L. Lampe, J. Newbury, and T. Swart (eds.), *Power Line Communications: Theory and Applications for Narrowband and Broadband Communications over Power Lines*, Chichester, United Kingdom: John Wiley & Sons, 2010.

[6] Lampe, L., *Power Line Communications: Principles, Standards and Applications from Multimedia to Smart Grid*, Chichester, United Kingdom: John Wiley & Sons, 2016.

[7] Tonello, A. M., and F. Versolatto, "Bottom-Up Statistical PLC Channel Modeling, Part I: Random Topology Model and Efficient Transfer Function Computation," *IEEE Transactions on Power Delivery*, Vol. 26, No. 2, April 2011, pp. 891–898.

[8] Canete, F. J., J. A. Cortes, L. Diez, and J. T. Entrambasaguas, "A Channel Model Proposal for Indoor Power Line Communications," *IEEE Communications Magazine*, Vol. 49, No. 12, 2011, pp. 166–174.

[9] Banwell, T., and S. Galli, "A Novel Approach to the Modeling of the Indoor Power Line Channel, Part I: Circuit Analysis and Companion Model," *IEEE Transactions on Power Delivery*, Vol. 20, No. 2, 2005, pp. 655–663.

[10] Galli, S., and T. C. Banwell, "A Deterministic Frequency-Domain Model for the Indoor Power Line Transfer Function," *IEEE Journal on Selected Areas in Communications*, Vol. 24, No. 7, 2006, pp. 1304–1316.

[11] Zimmermann, M., and K. Dostert, "A Multipath Model for the Powerline Channel," *IEEE Transactions on Communications*, Vol. 50, No. 4, 2002, pp. 553–559.

[12] Guzelgoz, S., H. B. Celebi, and H. Arslan, "Statistical Characterization of the Paths in Multipath PLC Channels," *IEEE Transactions on Power Delivery*, Vol. 26, No. 1, 2011, pp. 181–187.

[13] Tonello, A. M., F. Versolatto, B. Bejar, and S. Zazo, "A Fitting Algorithm for Random Modeling the PLC Channel," *IEEE Transactions on Power Delivery*, Vol. 27, No. 3, 2012, pp. 1477–1484.

[14] Paul, C. R., *Analysis of Multiconductor Transmission Lines*, Hoboken, NJ: John Wiley & Sons, 2008.

[15] Galli, S., and T. Banwell, "A Novel Approach to the Modeling of the Indoor Power Line Channel–Part II: Transfer Function and Its Properties," *IEEE Transactions on Power Delivery*, Vol. 20, No. 3, 2005, pp. 1869–1878.

[16] Babic, M., M. Hagenau, K. Dostert, and J. Bausch, "Theoretical Postulation of PLC Channel Model," Open PLC European Research Alliance (OPERA), 2005.

[17] Tlich, M., A. Zeddam, F. Moulin, and F. Gauthier, "Indoor Power-Line Communications Channel Characterization Up to 100 mHz, Part I: One-Parameter Deterministic Model," *IEEE Transactions on Power Delivery*, Vol. 23, No. 3, July 2008, pp. 1392–1401.

[18] Ahmed, M. O., and L. Lampe, "Power Line Network Topology Inference Using Frequency Domain Reflectometry," *Proc. IEEE International Conference on Communications (ICC)*, Ottawa, Canada, 2012, pp. 3419–3423.

[19] Ahmed, M. O., and L. Lampe, "Parametric and Nonparametric Methods for Power Line Network Topology Inference," in *Proc. IEEE International Symposium on Power Line Communications and Its Applications (ISPLC)*, Beijing, China, 2012, pp. 274–279.

[20] Zhang, C., X. Zhu, Y. Huang, and G. Liu, "High-Resolution and Low Complexity Dynamic Topology Estimation for PLC Networks Assisted by Impulsive Noise Source Detection," *IET Communications*, Vol. 10, No. 4, 2016, pp. 443–451.

[21] Passerini F., and A. M. Tonello, "Power Line Network Topology Identification Using Admittance Measurements and Total Least Squares Estimation," in *Proc. IEEE International Conference on Communications (ICC)*, Paris, France, 2017, pp. 1–6.

[22] Tonello, A. M., S. D'Alessandro, and L. Lampe, "Cyclic Prefix Design and Allocation in Bit-Loaded OFDM over Power Line Communication Channels," *IEEE Transactions on Communications*, Vol. 58, No. 11, 2010, pp. 3265–3276.

[23] D'Alessandro, S., A. M. Tonello, and L. Lampe, "Adaptive Pulse-Shaped OFDM with Application to In-Home Power Line Communications," *Telecommunication Systems*, Vol. 51, No. 1, 2012, pp. 3–13.

[24] Vo, T. N., K. Amis, T. Chonavel, and P. Siohan, "Achievable Throughput Optimization in OFDM Systems in the Presence of Interference and Its Application to Power Line Networks," *IEEE transactions on Communications*, Vol. 62, No. 5, 2014, pp. 1704–1715.

[25] Lampe, L., "Bursty Impulse Noise Detection by Compressed Sensing," in *Proc. IEEE International Symposium on Power Line Communications and Its Applications (ISPLC)*, Udine, Italy, 2011, pp. 29–34.

[26] Nassar, M., J. Lin, Y. Mortazavi, A. Dabak, I. H. Kim, and B. L. Evans, "Local Utility Power Line Communications in the 3–500 kHz band: Channel Impairments, Noise, and Standards," *IEEE Signal Processing Magazine*, Vol. 29, No. 5, 2012, pp. 116–127.

[27] Juwono, F. H., Q. Guo, D. Huang, and K. P. Wong, "Deep Clipping for Impulsive Noise Mitigation in OFDM-Based Power-Line Communications," *IEEE Transactions on Power Delivery*, Vol. 29, No. 3, 2014, pp. 1335–1343.

[28] Chien, Y.- R., "Iterative Channel Estimation and Impulsive Noise Mitigation Algorithm for OFDM-Based Receivers with Application to Power-Line Communications," *IEEE Transactions on Power Delivery*, Vol. 30, No. 6, 2015, pp. 2435–2442.

Part II:

Hybrid VHF TV White Space-Broadband Power Line Communication for Indoor High Speed IoT Networks

4

High-Speed Industrial IoT Networks: Gap between Existing Technologies and New Challenges

4.1 Introduction

In the last decade, remarkable growth was observed in the human population. This growth forced a rapid increase in the demand for diverse life sectors. The energy and communication sectors are considered among the top vital life sectors for the human population. The wide growth in human energy consumption, particularly in the industry sector, has driven researchers to focus on energy production from renewable sources [1] in the face of high scarcity of fossil fuels. On the other hand, growth in the human population has been accompanied by growth in the technological sector, information systems, and social media networking. The latter growth has put a huge pressure on the current communication systems that support those technologies. Hence, academics have focused their research on exploring new communication technologies or adapting current communication technologies to the existing challenges.

In a world with a growing demand for everything, the term "smart" is introduced in the energy and communication sectors to represent a new generation of technology, where it can adapt itself according to the customer requirement or the existing challenges [2]. For example, in smart grid, energy production, transmission, and consumption are bonded together through a good sensor network [3]. This sensor network provides the necessary information to generate exactly the required energy for consumption according to the conditions of the renewable energy sources. Regarding information systems and networking, this sensor network also provides information about customer networking activity, which helps the communication networks to better serve their customers more efficiently.

However, thousands of sensors inside this big sensor network can use a common platform that facilitates sharing their information, which is the Internet of Things (IoT). In the IoT network, different sensors with different communication technologies share their information for better overall network performance. However, the installation of such sensors in both energy and communication sectors can lead to more pressure on an already exhausted communication network that suffers from growing challenges in data speed and bandwidth. To better understand those challenges, the rest of this chapter will focus on describing the communication network in both industrial and residential environments and the need for novel communication technologies to address their requirements.

4.2 IoT Network in Industrial Areas

Operation and maintenance (O&M) of machines and electric equipment in an industrial environment is a vital component in any industry. For example, O&M of wind farms represent 15% of the total cost of energy production [4]. This number is expected to increase in the case of offshore wind farms.

Machine monitoring does not aim to only maintain normal operation; however, it predicts the possibility of its failure in an early stage. This greatly improves the replacement of different parts in the machine at correct time intervals and also decreases the visits of the trained personnel to offshore areas and hence, creates a consequent decrease in overall maintenance cost. As a result, machine monitoring can include, for example, rotation speed, vibration, thermal image, and water leakage. The sensor data for each machine is fed back to a central processing unit for further analysis to decide if there is any part in the machine that needs either a repair or replacement.

Different sensor networks can be used for machine monitoring. Sensor networks are classified into wired or wireless networks. However, wired sensor network can incur high expenses regarding cable installation and maintenance in the industrial areas. Also, the weight of those cables can represent a problem in some machine monitoring applications, as in the case of gas turbine engines used in the jet plane industry. Moreover, the process of installing those cables can be difficult in other industrial areas like inside a wind turbine tower.

WSNs have been considered as an alternative solution that can possibly replace wired sensor networks [5]. The main problem in the WSN is the high scattering communication environment that can degrade communication quality. As well, most of the machines rely on different types of high-current power supplies, which can cause electromagnetic interference to the communications, especially during switching instances as in the case of converter stations in wind farms.

In the energy sector, monitoring the power grid is an essential element for efficient energy consumption and production. PLC offers a cost-effective solution in this context [6]. PLC does not need extra installation of cables or communication platforms, and it provides good access to offshore areas using communications over both medium- and high-voltage power line cables. All these advantages make the PLC sensor network a better candidate in monitoring power grids compared to wireless sensor networks.

4.3 IoT Networks in Urban Areas

We mentioned earlier that the "smart" term is introduced in the energy sector to refer to the connection between energy production and consumption. "Smart" is also introduced in the urban environment to refer to a central automation system that can better manage different tasks with high speed and efficiency. For example, different appliances inside a smart home are connected to a central system that can better manage their consumption of the energy to either use less electricity or use the electricity at cheaper time tariffs [7]. This requires different sensors to be installed around the home and connected via an efficient sensor network.

In an outdoor urban environment, smart applications include smart traffic systems that organize car flow based on the intensity of a traffic jam [8]. As well, smart buildings have different sensors installed to monitor different levels of stress and vibration across the building, especially in bridges [9]. These monitoring services can guarantee high life-safety standards, which cannot be achieved today using nonautomated systems. However, these smart systems require good communication infrastructure for sensor data exchange.

4.4 IoT Networks in Agricultural Areas

Precision agriculture represents an automated management system that combines information and communication systems with agriculture tools for better agriculture production [10]. The system is completely adaptive to different weather and crop-growing conditions. This is achieved through different sensors that are installed to monitor levels of moisture in the fruits or soil, temperature, wind speed, or any other weather condition. The information collected can help determine an efficient irrigation schedule and monitor crop health and growth stages. Different topologies are developed for sensor networks to be deployed in agricultural areas whether they are terrestrial or underground networks.

4.5 IoT Heterogeneous Networks

To satisfy the requirements of IoT network applications in various fields, heterogeneous communication networks have been proposed as a promising solution. The revolutionary development in communication technologies and protocols that took place in the last decade assists information exchange with high speed and large bandwidth. This creates a reliable backhaul heterogeneous network that can support IoT services, which encompass connecting different objects and sensors with different communication technologies into one network [11,12]. TVWS communication and BPLC are two promising technologies that introduce cost-effective solutions for high-speed applications, especially in the indoor environment [13–16]. The two technologies have been recently introduced to support machine-to-machine services in IoT network [17–19]. However, TVWS and BPLC share the VHF band, which can introduce potential interference if they are integrated into the same IoT network. This interference can affect the achievable throughput and coverage distance, which drives our need for deeper study into the regulatory standards of both technologies to achieve better cooperation and integration into an IoT network.

Cognitive access of the TVWS spectrum was recommended by the FCC in May 2004 [20]. In December 2008, the FCC issued a second report [21] to regulate TVWS cognitive communication. The report regulates the transmission power limits over the SUs in order to protect the PUs from interference. SUs must acquire sufficient information about PU activity before utilizing the TV channel. Hence, the SU must be connected to a geolocation database and/or have high sensing capabilities. The geolocation database has a detailed temporal and geographical map for the availability of PU TV channels.

BPLC has drawn researchers' attention in the last decade as a promising technology for high-speed indoor applications. However, power line cables have not been designed or shielded to carry communication signals, which causes communication signal power loss due to electromagnetic radiation [22,23]. This radiation can cause severe interference to surrounding wireless services, which share the same band, in particular, TV and radio wireless services in the VHF band [24–26]. Consequently, EMC regulations limit the PSD of the BPLC transmitter to avoid interference with existing wireless services. For example, the standard of the Comite European de Normalisation Electrotechnique (CENELEC) [16] limits the PSD of BPLC signal to −55 dBm/Hz in the HF band between 1.8 MHz and 30 MHz, and −80 dBm/Hz in the VHF band between 30 MHz and 100 MHz. The 25-dB difference in the allowed PSD between the HF and the VHF band yields a significant degradation in BPLC achievable throughput and coverage distance in the VHF band compared to the HF band.

BPLC and TVWS are regarded as promising candidates for indoor broadband applications of the IoT. However, they share the access to the VHF band, which could cause harmful interference and performance degradation to each other. In this book, a TVWS-regulated BPLC system is proposed for point-to-multipoint downlink communication, which integrates the requirement of primary user sensing and the permissible transmission PSD for TVWS users into the BPLC standard in terms of VHF band access. This integration guarantees minimum interference level between TVWS and BPLC and allows higher transmission PSD for BPLC users in VHF, and hence higher capacity and coverage for BPLC.

4.6 Summary

In this chapter, IoT network applications in diverse life fields have been presented. There is a rapid growing need for accurate monitoring of the energy consumption by humans and machines. This monitoring can assist in more energy-efficient use in the presence of scarce energy resources. The increase in health monitoring applications has lead to a corresponding increase in the number of sensors that are connected together via different communication networks. TVWS and BPLC communication networks appear to be efficient communication technologies that are being required to be integrated in future IoT networks. However, both solutions access the same frequency band, which can create possible interference between both technologies. In the next chapters we will investigate more into this problem and find out how can we integrate both solutions in one IoT network with minimum interference and maximum efficiency.

References

[1] Owusu, P. A., and S. Asumadu-Sarkodie, "A Review of Renewable Energy Sources, Sustainability Issues and Climate Change Mitigation," *Cogent Engineering*, Vol. 3, No. 1, 2016, pp. 1167990.

[2] Haider, T. H., O. H. See, and W. Elmenreich, "A Review of Residential Demand Response of Smart Grid," *Renewable and Sustainable Energy Reviews*, Vol. 59, 2016, pp. 166–178.

[3] Fadel, E., V. C. Gungor, L. Nassef, et al, "A Survey on Wireless Sensor Networks for Smart Grid," *Computer Communications*, Vol. 71, 2015, pp. 22–33.

[4] Martin, R., I. Lazakis, S. Barbouchi, and L. Johanning, "Sensitivity Analysis of Offshore Wind Farm Operation and Maintenance Cost and Availability," *Renewable Energy*, Vol. 85, 2016, pp. 1226–1236.

[5] Sherazi, H. H. R., G. Piro, L. A. Grieco, and G. Boggia, "When Renewable Energy Meets LORA: A Feasibility Analysis on Cable-Less Deployments," *IEEE Internet of Things Journal*, Vol. 5, No. 6, 2018, pp. 5097–5108.

[6] Bumiller, G., L. Lampe, and H. Hrasnica, "Power Line Communication Networks for Large-Scale Control And Automation Systems," *IEEE Communications Magazine*, Vol. 48, No. 4, 2010, pp. 106–113.

[7] Gaikwad, P. P., J. P. Gabhane, and S. S. Golait, "A Survey Based on Smart Homes System Using Internet-of-Things," in *2015 International Conference on Computation of Power, Energy, Information and Communication (ICCPEIC)*, IEEE, 2015, pp. 0330–0335.

[8] Rizwan, P., K. Suresh, and M. Rajasekhara Babu, "Real-Time Smart Traffic Management System for Smart Cities by Using Internet of Things and Big Data," in *2016 International Conference on Emerging Technological Trends (ICETT)*, IEEE, 2016, pp. 1–7.

[9] Moreu, F., R. E. Kim, and B. F. Spencer Jr., "Railroad Bridge Monitoring Using Wireless Smart Sensors," *Structural Control and Health Monitoring*, Vol. 24, No. 2, 2017, p. e1863.

[10] Srbinovska, M., C. Gavrovski, V. Dimcev, A. Krkoleva, and V. Borozan, "Environmental Parameters Monitoring in Precision Agriculture Using Wireless Sensor Networks," *Journal of Cleaner Production*, Vol. 88, 2015, pp. 297–307.

[11] Al-Fuqaha, A., M. Guizani, M. Mohammadi, M. Aledhari, and M. Ayyash, "Internet of Things: A Survey on Enabling Technologies, Protocols, and Applications," *IEEE Communications Surveys & Tutorials*, Vol. 17, No. 4, June 2015, pp. 2347–2376.

[12] Xu, L. D., W. He, and S. Li, "Internet of Things in Industries: A Survey," *IEEE Transactions on Industrial Informatics*, Vol. 10, No. 4, January 2014, pp. 2233–2243.

[13] Kang, K.-M., and J. Park, "A New Scheme for Compliance with TV White Space Regulations using Wi-Fi Modules in a Cognitive Radio System," *IEEE Transactions on Consumer Electronics*, Vol. 60, No. 4, November 2014, pp. 567–573.

[14] Wang, H., Y. Qian, and H. Sharif, "Multimedia Communications over Cognitive Radio Networks for Smart Grid Applications," *IEEE Wireless Communications*, Vol. 20, No. 4, August 2013, pp. 125–132.

[15] Rehman, M. U., S. Wang, Y. Liu, S. Chen, X. Chen, and C. G. Parini, "Achieving High Data Rate in Multiband-OFDM UWB over Power-Line Communication System," *IEEE Transactions on Power Delivery*, Vol. 27, No. 3, July 2012, pp. 1172–1177.

[16] Ferreira, H. C., L. Lampe, J. Newbury, and T. Swart, (eds.), *Power Line Communications: Theory and Applications for Narrowband and Broadband Communications over Power Lines*, Chichester, United Kingdom: John Wiley & Sons, Inc., 2010.

[17] Aijaz, A., and A. H. Aghvami, "Cognitive Machine-to-Machine Communications for Internet-of-Things: A Protocol Stack Perspective," *IEEE Internet of Things Journal*, Vol. 2, No. 2, April 2015, pp. 103–112.

[18] Hsieh, H.-C., and C.-H. Lai, "Internet of Things Architecture Based on Integrated PLC and 3G Communication Networks," in *IEEE 17th International Conference on Parallel and Distributed Systems (ICPADS)*, December 2011, Tainan, Taiwan, pp. 853–856.

[19] Cano, C., A. Pittolo, D. Malone, L. Lampe, A. M. Tonello, and A. G. Dabak, "State of the Art in Power Line Communications: From the Applications to the Medium," *IEEE Journal on Selected Areas in Communications*, Vol. 34, No. 7, July 2016, pp. 1935–1952.

[20] Federal Communications Commission, Notice of Proposed Rule Making, Document 04–113, May 2004.

[21] Federal Communications Commission, Second Report and Order and Memorandum Opinion and Order In the Matter of Unlicensed Operation in the TV Broadcast Bands, Additional Spectrum for Unlicensed Devices Below 900 MHz and in the 3 GHz Band, Document 08-260, November 2008.

[22] Gebhardt, M., F. Weinmann, and K. Dostert, "Physical and Regulatory Constraints for Communication over the Power Supply Grid," *IEEE Communications Magazine*, Vol. 41, No. 5, May 2003, pp. 84–90.

[23] Oliveira, T., F. Andrade, A. Picorone, H. Latchman, S. Netto, and M. Ribeiro, "Characterization of Hybrid Communication Channel in Indoor Scenario," *Journal of Communication and Information Systems*, Vol. 31(1), September 2016.

[24] Berger, L. T., A. Schwager, P. Pagani, and D. Schneider, *MIMO Power Line Communications: Narrow and Broadband Standards, EMC, and Advanced Processing*, Boca Raton, FL: CRC Press, 2014.

[25] Pagani, P., R. Razafferson, A. Zeddam, et al., "Electromagnetic Compatibility for Power Line Communications," in *Proc. IEEE 21st International Symposium on Personal Indoor and Mobile Radio Communications (PIMRC)*, September 2010, Istanbul, Turkey, pp. 2799–2804.

[26] Ronkainen, T., R. Vuohtoniemi, and J.-P. Makela, "Radiated Interference of High Frequency Broadband Power Line Communications," *International Symposium on Electromagnetic Compatibility (EMC Europe)*, September 2014, Gothenburg, Sweden, pp. 555–559.

5

Cross Talk between TV White Space and Power Line Communication

5.1 Introduction

In this chapter, the BPLC and the TVWS channels are modeled along with the cross-talk channel, and a general statistics based path loss mapping (GSBPLM) approach is proposed for BPLC.

The chapter is organized as follows. Section 5.2 presents the proposed GSBPLM criteria to map the path loss for each PLC channel class. Also, the section presents a proposed simplification method adopted in GSBPLM for BPLC channel transfer function computation referred to as the gain rollback (GRB) method. Furthermore, the simulated GSBPLM path loss is compared to the measured one, and a comparison between the PLC channel path loss and the wireless channel path loss is presented as well. In Section 5.3, the TVWS BPLC cross-talk channel model and measurements are shown. In Section 5.4, a summary of the chapter is presented.

5.2 GSBPLM PLC Path Loss Mapping

The power line channel is divided into nine classes [1] based on achievable channel capacity (more specifically Shannon's capacity), where channel 1 is the channel with the least capacity and channel 9 is the channel with the highest capacity. In other words, channel 1 represents the worst BPLC channel gain conditions while channel 9 represents the best BPLC channel conditions. The main target of the work is to map the path loss of the PLC channel with all its different scenarios. The work is different from [2] and [3] in the following aspects. First, a novel method is proposed to map the path loss and the coverage distance of each class of the PLC channels in [1] using a statistical approach. A new methodology has been followed in mapping and measuring the power line signal coverage distance. The new methodology depends introducing the *electric* coverage distance concept rather than the *geometric* one which was previously used in [3]. This new concept increases the accuracy of the simulated path loss map and decreases the difference between the simulations and measurements. The proposed path loss map is more general than the previous work in [2] and [3] and is applicable for different topologies and environments within each channel class. Since the nine PLC channel classes [1] represent most of the existing power line topologies based practical measurements, the topology that represents each PLC class channel needs to be modeled clearly. Building up nine corresponding random topology generators is proposed, which include all the possible scenarios of the PLC channel environments, leading to more general and practical path loss mapping. Second, a new GRB approach is used to reduce the computational complexity of the channel transfer function between two arbitrary nodes in a complex power line network. The proposed GRB method is more general and less complex than that in [4], by deriving one general expression for every branch gain. The method takes all the complex situations of the power line network into consideration, leading to more practical results. Another essential advantage of the GRB method is the effect of the adjacent nodes being taken into consideration, which yields more accurate results than the other bottom-up methods [5,6].

Simulation and measurement results show that the proposed topology generators are able to simulate the nine PLC channel classes effectively. The average channel gain curve of every class of topology generator matches the fitting curve of the corresponding PLC channel class in [1]. The path loss curves of the nine PLC channel classes are generated at different frequencies in the VHF band, and are compared to their counterparts of wireless communications to show the feasibility of the cooperation between the BPLC and wireless communications in indoor environments.

Hence, in order to achieve our target, the PLC channel response has to be simulated for every class of the PLC channels presented in [1]. In [7], a statistical model was presented for building up a power line topology. However, this model did not identify the class of the generated topology, and therefore is inadequate to represent all the different measurement environments of the PLC channels. In this section, the aim is to statistically model the power line topology representing each PLC channel class. Different topology parameters are presented, which are adjustable to build up nine classes of the power line topology generators. Each topology class generator has an average channel gain that represents the gain of the corresponding PLC channel class presented in [1]. The adjustable parameters can be summarized into three categories described below.

5.2.1 Topology Layout

The number of branches has a great impact on the overall channel gain degradation. Hence, controlling the number of branches in each power line topology will define the overall topology class. This is achieved by controlling the area of the small clusters building up the whole topology and the number of the electric outlets within each cluster. In the proposed approach for simplicity, the topology area A_f is assumed to be divided into square-shape clusters of identical area A_{cn} where n denotes the index of the topology class. A_{cn} is assumed to be uniformly distributed over $[A_m, A_M]$, where A_m and A_M represent the minimum and maximum values for each topology class.

5.2.2 Load Distribution

Controlling the model of the load distribution in each topology will determine the class of this topology. The adopted model for the load distribution in the proposed approach classifies the load impedance connected to every outlet into three groups: frequency dependent, frequency independent and open circuit (i.e., no load connected). Let $Z(\omega)$ denotes the impedance of the frequency dependent loads at frequency ω. It can be expressed as [8]

$$Z(\omega) = \frac{R_L}{1 + jQ(\frac{\omega}{\omega_0} - \frac{\omega_0}{\omega})} \tag{5.1}$$

where R_L is the load impedance value at resonance frequency, Q is the quality factor, and ω_0 is the angular resonance frequency. For the frequency independent loads $Z(\omega) = R_L$ for the whole band, where each load has different R_L value.

5.2.3 Cable Modeling

The power line cable model is essential in determining the class of the topology since it determines the cable attenuation factor and hence determines the overall channel attenuation. In this work, the cable is represented using the empirical relation of [5]: $\alpha(f) = p_{1n} + p_{2n}f + p_{3n}f^2 + p_{4n}f^3$, where $\alpha(f)$ is the cable attenuation factor at frequency f Hz, p_{1n}, p_{2n}, p_{3n}, and p_{4n} are the cable constants for the nth topology class.

5.2.4 GRB Method for Channel Transfer Function Computation

The adjacent nodes to the Tx and the Rx influence the overall channel transfer function. However, due to the complexity of power line network, it was difficult to take the effect of adjacent nodes accurately into consideration in the previous work, such as in the impedance carryback method of [7].

In this part, a GRB method is presented for computing the channel transfer function between two arbitrary nodes in a complex power line network. The

GRB method has an advantage of reduced computational complexity by breaking down the overall channel gain computation into smaller computation operations for each branch in the power line network using the time domain approach. The GRB method is also more general than [4], and the calculation of the adjacent nodes effect leads to more practical results.

The time domain approach was previously addressed in [4] to compute the PLC channel transfer function taking the adjacent node effect into consideration. However, the most significant drawback of this approach is the assumption that the obstructed (OBS) branches are terminated with terminal loads only. In contrast in the real power line networks the OBS branches can be terminated by either another branching node or terminal impedance. Hence, through this part a general formula will be derived for the OBS branch gain taking into consideration the practical case. Finally, a simplification method will be addressed to compute the overall channel transfer function between any two arbitrary nodes in the network using the new general formula derived for the OBS branch gain.

In Figure 5.1, the overall transfer function between the Tx and the Rx can be expressed using the following formula [4]:

$$H(f) = LOS\prod_{i=1}^{n}(OBS_i + 1)(A_L + 1)(A_R + 1) \tag{5.2}$$

where LOS denotes the line-of-sight path gain, OBS_i is the obstructed path gain of the ith branch, A_L and A_R are the sums of the multipath signal propagating on the left hand side of the Tx and the right hand side of the Rx, respectively, and n is the total number of the obstructed branches and paths coming out of the LOS direction. The LOS is given by the product of the transmission coefficients and the line attenuation factor as $\prod_{i=1}^{n}(T_i e^{-\gamma l})$, where T_i represents the transmission coefficient of the ith branch, γ is the propagation constant, and l is the branch

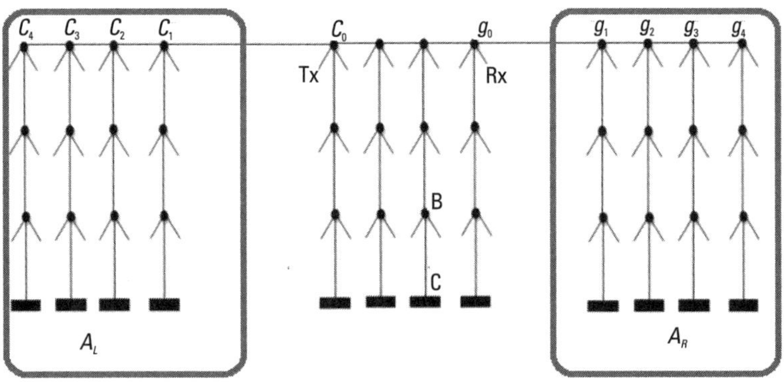

Figure 5.1 Power line real network with multiple branches and multiple adjacent nodes.

length. The $OBS_{q,s}$ for the qth node and the sth branch can be expressed as

$$OBS_{q,s} = \begin{cases} \frac{a_s}{1-r_s} & \text{if } (q+1) \text{ is a terminal node} \\ N_{q+1} + & \text{if } (q+1) \text{ is a branching node} \\ \quad M_{q+1}[\prod_{s=1}^{n}(OBS_{q+1,s}+1)-1] & \end{cases} \tag{5.3}$$

The derivation of (5.3) is shown in Appendix A using the adjacent node gain method in [4]. It can be deduced that the values of A_L and A_R in (5.2) can be considered as the overall equivalent OBS values for the two adjacent branches on the left-hand side of Tx and the right-hand side of the Rx, respectively, and hence (5.2) can be rewritten as

$$H(f) = LOS \prod_{i=1}^{n+2}(OBS_i + 1) \tag{5.4}$$

Consequently, the GRB method can be simplified in the following steps:

1. Determine the line of sight channel gain between the Tx and the Rx;
2. Evaluate the OBS gain of each branch diverting from the LOS path using the general formula (3);
3. Evaluate the overall channel transfer function using (4) taking the adjacent nodes into account.

5.2.5 Simulations versus Measurements

5.2.5.1 Simulation Setup

The adjustable parameters for building up each class of the PLC topology generators are addressed in Tables 5.1 and 5.2. The parameters are deduced using iterative procedure that minimizes the difference between the simulated

Table 5.1
Topology Generator Parameters

Channel Class	A_{mn} (m^2)	A_{Mn} (m^2)	Outlet Density ($\frac{outlet}{m^2}$)	ρ_{v1n}	ρ_{v2n}
1	15	45	0.05	0.4	0.1
2	26	45	0.035	0.4	0.1
3	35	47	0.03	0.5	0.05
4	45	55	0.025	0.6	0.08
5	60	70	0.025	0.8	0.08
6	90	100	0.01	0.8	0.08
7	190	200	0.015	0.8	0.2
8	450	480	0.02	0.8	0
9	450	480	0.0125	0.8	0.08

Table 5.2
Cable Parameters

Channel Class	p_{1n}	p_{2n}	p_{3n}	p_{4n}	Z_{cn} (Ω)
1	1.233e-03	-7.04e-10	1.05e-17	1.12e-26	120
2	5.2e-03	-1.77e-10	1.04e-17	-5.50e-26	122
3	5.2e-03	1.34e-10	3.66e-18	-1.63e-26	80
4	2.7e-03	1.23e-11	5.53e-18	-2.21e-26	80
5	1.8e-03	-7.58e-11	5.74e-18	-1.72e-26	80
6	8.7e-03	-5.32e-10	1.47e-17	-5.94e-26	80
7	4.1e-03	-3.67e-10	1.21e-17	-4.45e-26	80
8	3.2e-03	-1.64e-10	4.83e-18	-1.12e-26	20
9	2.7e-03	-9.65e-11	5.16e-19	7.76e-27	20

model and the fitting curve in [1]. The simulations are held to generate 400 PLC random topologies per each class topology generator. For every class topology, two arbitrary outlets are selected at random and separated by random distance to represent the Tx and the Rx, then the channel transfer function is computed between the two outlets using the GRB approach in Section 3.3.4. Consequently, the average transfer function is calculated for each class and the average channel gain is computed at different distances for each class topology generator at different frequencies.

In Table 5.1, the maximum cluster area A_M, the minimum cluster area A_m, the outlet density, the no load ratio ρ_{v1n} and the frequency dependent load ratio ρ_{v2n} are addressed for the nine classes of the random topology generators. The overall topology area A_f is fixed to 500 m^2.

Table 5.2 shows the characteristic impedance Z_{cn} and the coefficients p_{1n}, p_{2n}, p_{3n}, and p_{4n} of the empirical relation representing the power line cable for each class.

For the frequency dependent loads, the R, Q, and ω_0 are uniformly distributed in the ranges [200, 1,800] Ω, [5, 20], and [2,100] MHz, respectively. For the simulations, β, α and n in wireless channel model in (2.1) are assumed to be 2.63, 0.65, and 1.5, respectively, for the indoor environment as in [9]. *FAF* in (2.1) is assumed to be zero for simplicity, as the simulated area is assumed to be on the same floor.

5.2.5.2 Measurement Setup

To validate the simulation results, measurements are held for the power line channel path loss. The measurements were different from the previous ones by introducing the concept of the *electric* distance between the outlets. The electric distance can be defined as the distance along the power line cable connecting the

Tx and the Rx. However, the *geometric* distance can be defined as the Euclidean norm between the coordinates of the Tx and the Rx as in [3] regardless of the power line cable length. Hence, the new concept increases the accuracy of the path loss map measured as the electric distance is the effective factor in the signal attenuation and not the geometric one. The measurements were held in the Electrical Engineering building of Liverpool University. The Building is divided into six floors. Each floor is subdivided into several rooms where each group of rooms are connected to a dedicated distribution box. The outlets inside each room are connected together via a single phase 3-conductor power line cable.

The measurements have been carried out using both the signal generator (SG) and the network analyzer (NA). For the small distances (i.e., <15m) and the calibration process, the network analyzer was used for measuring the channel response by knowing the S-parameters. However, for the long distances (i.e., >15 m) it was more convenient (due to safety issues regarding extending cables between different rooms) to use the signal generator as a Tx and the network analyzer (in the spectrum analyzer mode) as a Rx. Two broadband coupling circuits were used in the Tx and the Rx. The loss has been measured in the coupling circuits and the connection cables by measuring their S-parameters using the network analyser for calibration purpose. As the connection cables are short, the loss inside those cables was nearly 0 dB. For the coupling circuits, the loss inside each one was almost constant value of 5 dB in the frequency range 2 MHz–100 MHz. The measurements were carried out between the outlets belonging to the same phase and the same distribution box. The electric distance between each two outlets was calculated knowing the building topology. The measurements were done in different rooms to have an average path loss representation. The channel transfer function was calculated according to the measurement setup:

1. NA setup:
   $H(f)(dB) = NA(dB) + Coupling\ Circuit\ Loss\ (dB).$
2. SG-NA setup:
   $H(f)(dB) = NA(dB) - SG(dB) + Coupling\ Circuit\ Loss\ (dB).$

The distance between each two outlets is measured according to the outlet location:

1. If the outlets are inside the same room, the distance is measured along the power line cable connecting those two outlets;
2. If the outlets are in two different rooms, the distance is measured along the power line cable between each outlet and the distribution box.

In Figures 5.2 and 5.3, the coupling circuit and its electronic component structure are shown. The main role of the coupling circuit is to isolate the mains voltage and to couple the communication signal to/from the power line network. The isolation is completely achieved through the 100-pF capacitor and also the

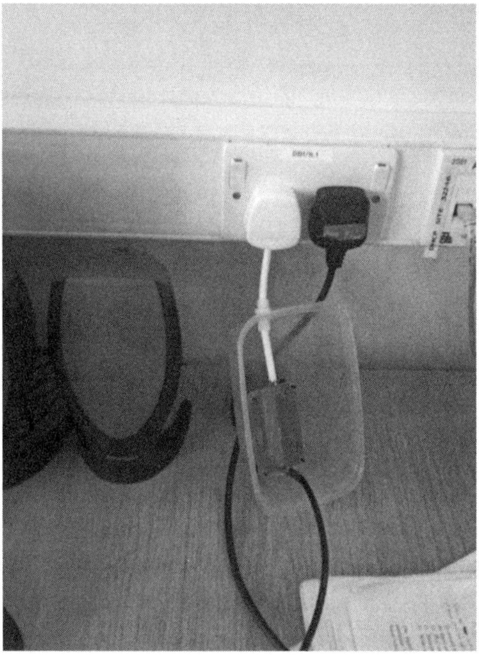

Figure 5.2 BPLC coupling circuit.

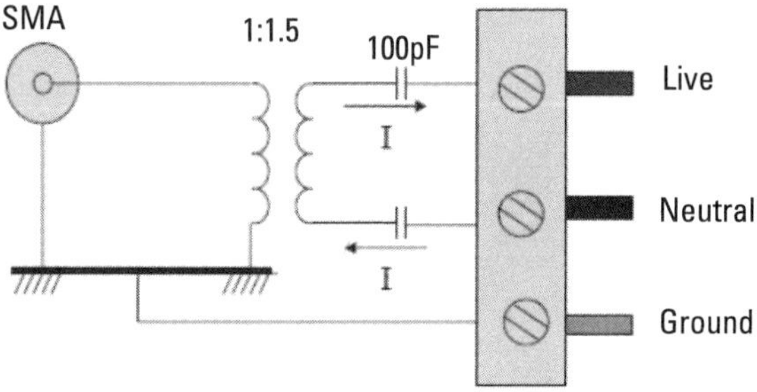

Figure 5.3 Schematic diagram of the BPLC coupling circuit [10].

inductive transformer. The inductive transformer physically isolates the mains voltage from any measurement device, which guarantees maximum protection. In Figures 5.4 and 5.5, the practical measurement setup for evaluating the BPLC channel gain is shown. In Figure 5.4, the RF signal generator sends the BPLC signal to the power line network via the coupling circuit, while in Figure 5.5, the network analyzer receives the signal through another coupling circuit. The network analyzer is adjusted to be in the spectrum analyzer mode to calculate the BPLC channel gain. Mapping the measured channel gain versus the distance between the Tx and Rx electric outlets, the path loss curve can be deduced. The measurements are held in rooms 221, 202, and 207 on the second floor of the

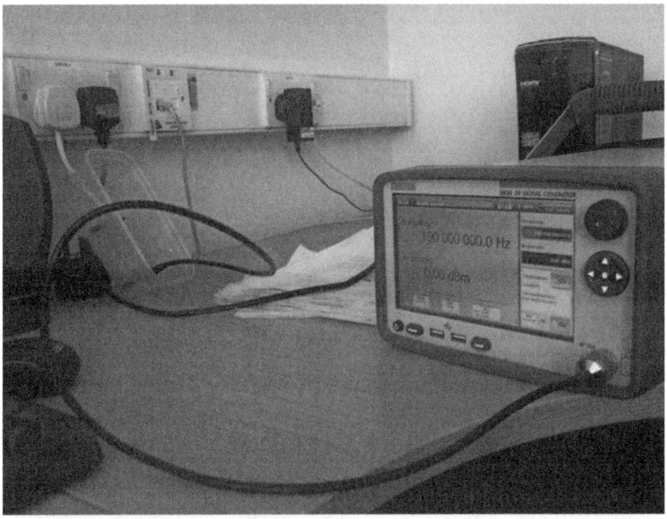

Figure 5.4 RF signal generator sends a single-tone frequency to the power line network.

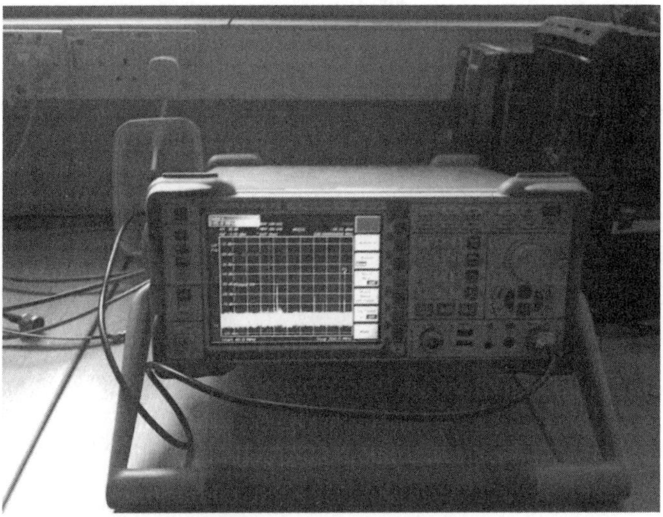

Figure 5.5 Network analyzer receives the power line signal.

Electrical Engineering-Building at the University of Liverpool. Since the distances between the outlets represent an important parameter in our measurements, the room maps are presented in the Figures 5.6, 5.7, and 5.8.

5.2.5.3 Simulated versus Measured PLC Channel Path Loss

Figures 5.9–5.10 show the path loss measured at frequencies 30 MHz and 100 MHz, respectively. Also in the figures the linear fitted curve deduced from

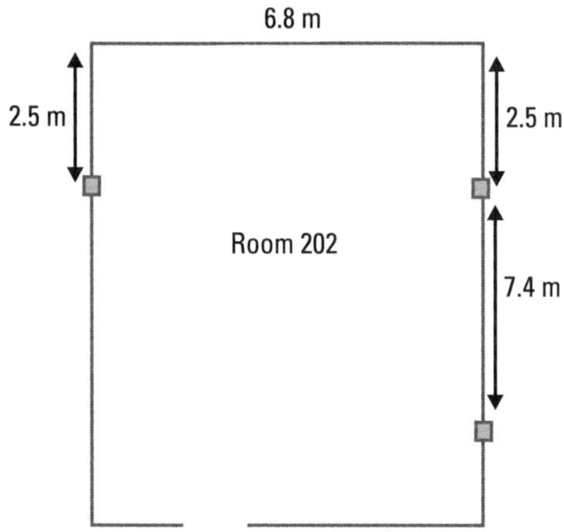

Figure 5.6 Outlet distribution map in room 202 in the Electrical Engineering Building at Liverpool University. The squares represent the electric outlets locations.

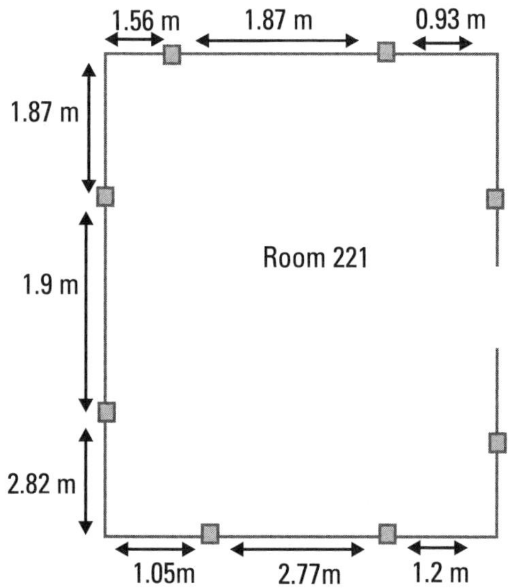

Figure 5.7 Outlet distribution map in room 221 in the Electrical Engineering Building at Liverpool University. The squares represent the electric outlets locations.

the measured data is compared with the corresponding simulated curve of class (1) channel using GSBPLM. The normalized mean square errors (NMSE) are calculated for both fitted curves. The NMSE is calculated according to the following equation:

$$NMSE = \frac{\| x_{meas} - x_{fit} \|^2}{\| x_{meas} - mean(x_{meas}) \|^2} \tag{5.5}$$

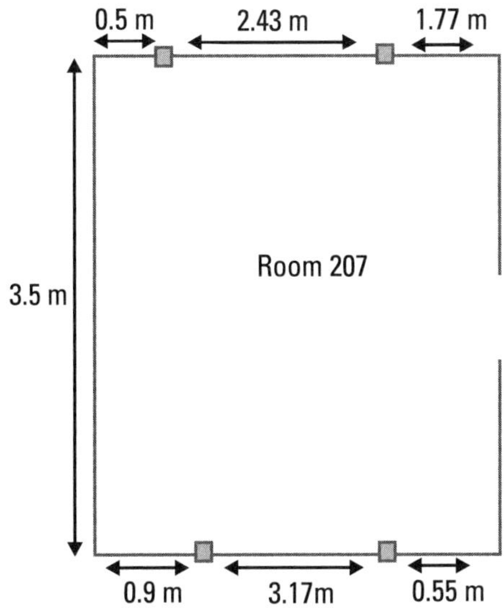

Figure 5.8 Outlet distribution map in room 207 in the Electrical Engineering Building at Liverpool University. The squares represent the electric outlets locations.

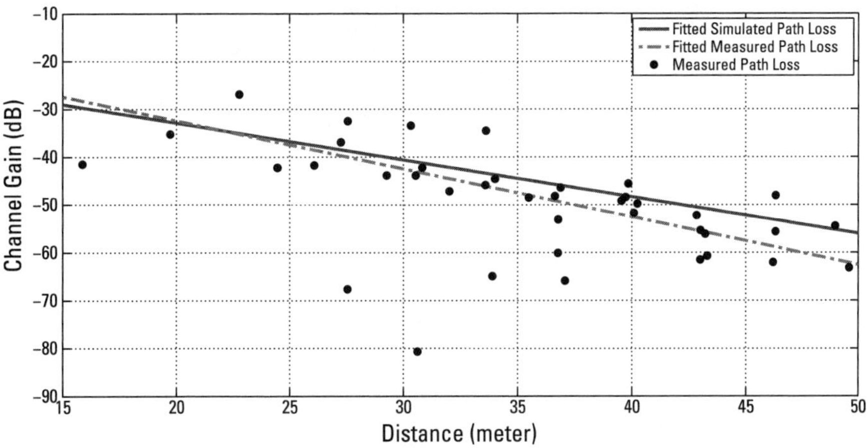

Figure 5.9 The measured, fitted, and simulated path loss curves of class 1 PLC channel at 30 MHz.

where x_{meas} and x_{fit} represent the measured and fitted data, respectively. It is found that the difference in the NMSE between both fitted curves is: 0.05 and 0.04 at frequencies 30 MHz and 100 MHz, respectively. Also, when computing the correlation coefficient between the measured and fitted data, the result is 0.857. This result proves that the simulated path loss agrees with the real measurement data that further proves the feasibility of the proposed path loss mapping approach.

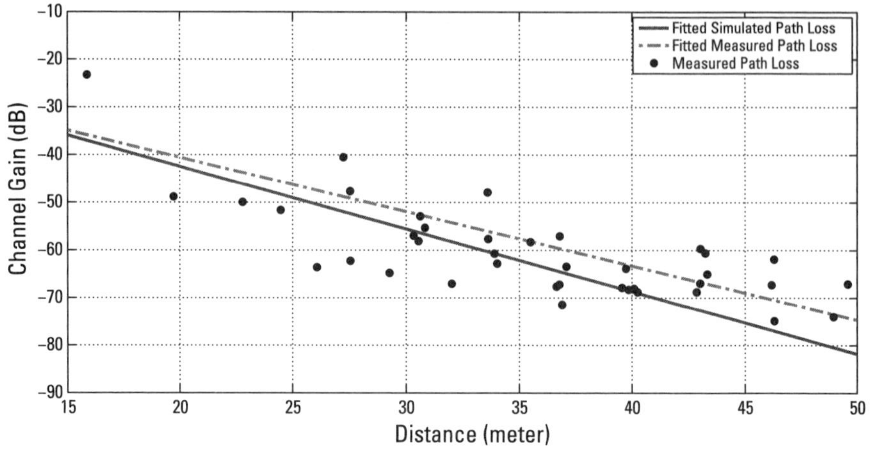

Figure 5.10 The measured, fitted, and simulated path loss curves of class 1 PLC channel at 100 MHz.

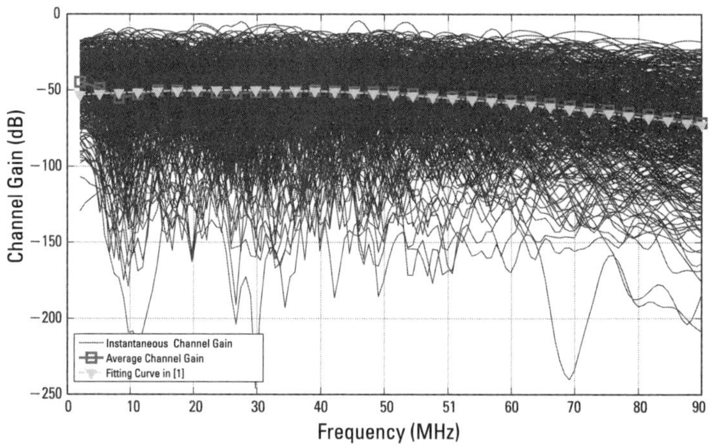

Figure 5.11 The transfer functions generated from class 1 topology generator.

5.2.5.4 Simulated Average PLC Channel Gain

In Figure 5.11, the channel transfer functions in the frequency range 2 MHz–100 MHz are plotted for the class (1) random topology generator. It is worth mentioning that the class 1 power line channel has been mostly measured and addressed in previous literature [3,11–13]; hence, its average channel path loss can be deduced as

$$PL_{\mathrm{BPLC}}(dB) = (a_1 f^2 + a_2 f + a_3)d + (b_1 f^2 + b_2 f + b_3) \tag{5.6}$$

where f is the frequency of operation in MHz, d is the electric separation distance in meters (i.e., the length of the separating power line cable), $a_1 = -1.14 \times 10^{-4}$,

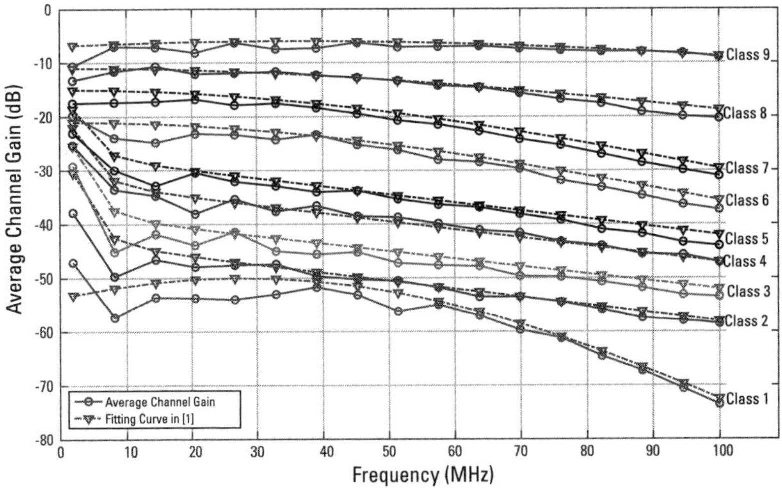

Figure 5.12 The average transfer functions of the nine classes generated by the nine topology generators and the fitting curves representing the nine classes.

$a_2 = 6.8 \times 10^{-3}$, $a_3 = -0.91$ and $b_1 = 2.97 \times 10^{-4}$, $b_2 = -0.061$, $b_3 = -15.0$. In Figure 5.12, the overall average transfer functions of the nine classes are plotted together with the corresponding fitting curves. The matching can be observed between the average transfer function generated and the fitting curve representing each class according to [1], which further proves the validity of the proposed topology generator in representing the corresponding class of the power line channels.

5.2.5.5 Simulated Path Loss of PLC versus Wireless Channel

In Figures 5.13–5.14, the path loss curves for the nine PLC classes are deduced from the nine topology generators within a coverage distance below 60m. The fitting curve for each class is drawn at frequencies 30 and 100 MHz. The GUPL model is also drawn to represent the path loss for the wireless channel in the indoor environment at the same frequencies and within the same coverage distance range. It is worth mentioning that the distance considered in the GUPL model is the geometric distance. However, the distance for PLC path loss is the electrical one. It can be easily concluded that the electrical distance is greater than or equal to the geometrical one. In Figures 5.13–5.14, it can be observed that for a coverage distance below 60m, the path losses of the PLC classes 1–5 together with the path loss of the wireless channel exceed 40 dB at different distances. Hence, the cooperation between the PLC and the wireless communication becomes feasible in the indoor environment, taking into consideration the difference between the electric and the geometric coverage distances. However, further study should be

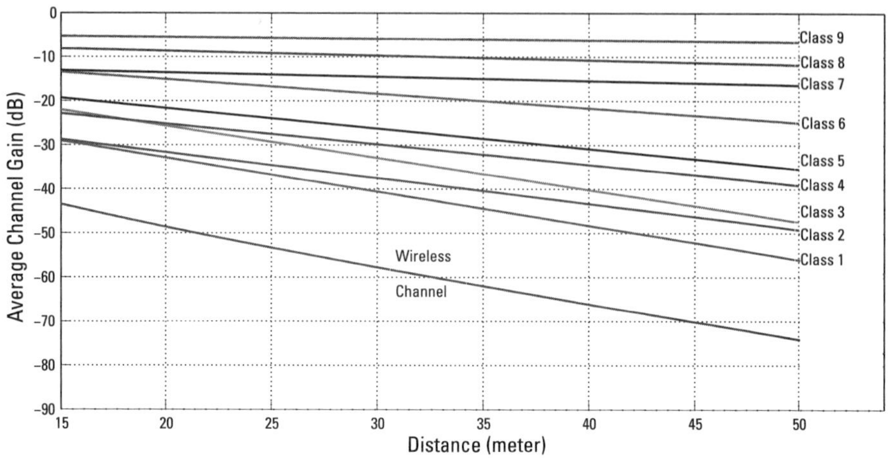

Figure 5.13 The deduced path loss curves of the nine classes of the power line channel and the path loss curve of the wireless channel at frequency 30 MHz.

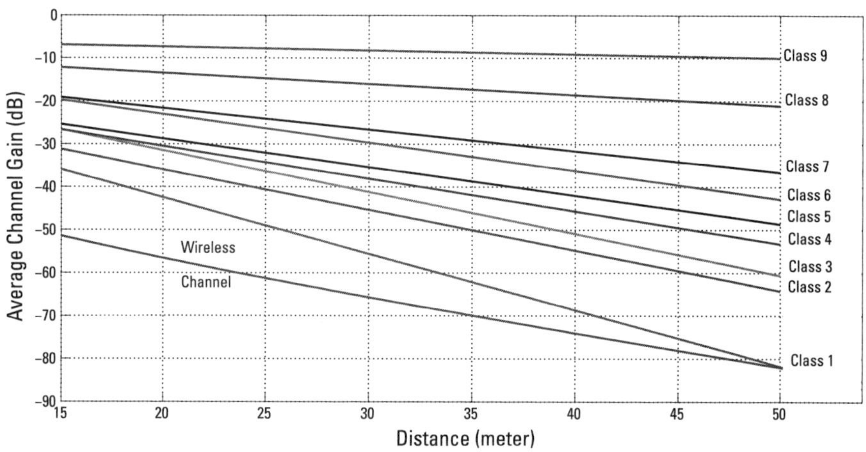

Figure 5.14 The deduced path loss curves of the nine classes of the power line channel and the path loss curve of the wireless channel at frequency 100 MHz.

considered for the other parameters affecting the cooperation between the BPLC and the wireless communications.

5.3 Modeling the Cross Talk between TVWS and BPLC Channels

As aforementioned, the TVWS wireless channel path loss is represented using the GUPL model in [14]. In this work, the indoor wireless channel is represented using 3-level model, which considers three superimposed effects: (1) Two path Rayleigh fading channel for the multipath effect [15], with RMS delay spread of

6 μs, (2) Shadowing effect, and (3) Path loss effect. Both shadowing and path loss effects are modeled using the GUPL model in (2.1).

For the cross talk between the TVWS and BPLC channels, the near field coupling between broadband power line and high frequency communication systems has been addressed in [16]. It depends on several factors like the cable properties, the coupling circuit and the network asymmetric loads.

In this section, the measurements done in the previous sections for the BPLC channel are continued to include the cross talk between the BPLC and the wireless channel. The aim of the measurements is to model the path loss of the cross-talk channel and compare it with the path loss of the corresponding BPLC and TVWS channels. The measurements are held inside the labs of the electrical engineering and electronics department of Liverpool University. The Tx and the Rx are located at the same floor. The channel between any two power line couplers is represented by the network of the power line cables, while the channel between any two wireless antennas is the radio propagation channel. The path loss is measured using a RF signal generator at the Tx and spectrum analyzer at the Rx. The frequency band of operation is 84 MHz–200 MHz, but the frequency modulation (FM) band from 88 MHz–110 MHz is avoided due to interference from the FM radio station. The path loss is measured for three different channels: (1) the channel between two power line coupling circuits, referred to as h_{11}; (2) the channel between a coupling circuit at the Tx and wireless antenna at the Rx, referred to as h_{12}, and (3) the channel between wireless antenna at the Tx and power line coupling circuit at the Rx, referred to as h_{21}.

In Figure 5.15, the gain of the cross-talk channel is shown versus the gain of the BPLC channel. The channel gain is measured at four frequencies in the band 84 MHz–200 MHz, which are 84 MHz, 110 MHz, 140 MHz, and 190 MHz. The gain is measured for each frequency at different coverage distances to model the path loss of the channel. The measurement results show convergence between the gain of the BPLC channel and the cross-talk channel, since the slope α of the fitted channel gain takes the values of -2.7, -1.8, and -1.4 for the h_{11}, h_{12}, and h_{21}, respectively. The results also show that the cross-talk channel gain is below the BPLC channel gain for small coverage distance; however, for long coverage distances the cross-talk channel gain outperforms the corresponding BPLC channel gain. This is due to the dependence of the cross-talk channel gain mainly on the geometric separation distance between the Tx and the Rx. However, BPLC channel gain is mainly dependent on the electric separation distance, which is the length of the power line cable connecting the Tx to the Rx. It is known that the electric distance is longer than the geometric distance as reported in [12]. Consequently, the average gain of the cross-talk channel can be concluded to be comparable to the average gain of the main BPLC or wireless TVWS channels. This agrees with the results addressed in [17–21], where the cross talk between the BPLC and the wireless channel was exploited to enhance

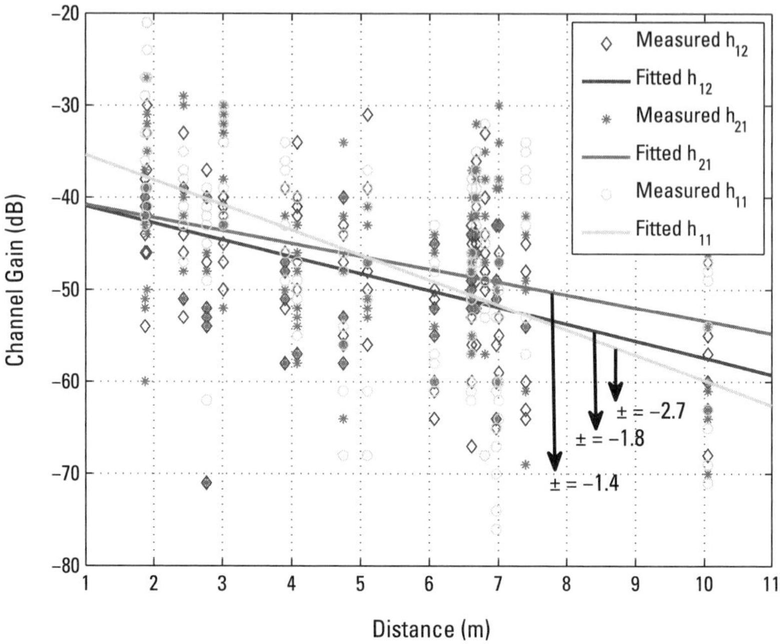

Figure 5.15 Cross-talk channel compared to the BPLC channel.

the reception of the BPLC signal. Hence, to simplify the analysis, the BPLC (Tx)-TVWS (Rx) channel can be represented as the sum of the BPLC-BPLC loss ($PL_{\text{B}-\text{B}}$) and the TVWS BPLC (PL_{c}) coupling loss (The average value of the coupling loss considered in this chapter is taken from the results of the carried-out measurements, which is 2 dB). As well, the TVWS (Tx)-BPLC (Rx) channel is represented as the sum of the TVWS-TVWS ($PL_{\text{T}-\text{T}}$) loss and the TVWS-BPLC coupling loss. This is illustrated in the following equation:

$$\begin{aligned} PL_{\text{BPLC}-\text{TVWS}}(dB) &= PL_{\text{B}-\text{B}}(dB) + PL_{\text{c}}(dB) \\ PL_{\text{TVWS}-\text{BPLC}}(dB) &= PL_{\text{T}-\text{T}}(dB) + PL_{\text{c}}(dB) \end{aligned} \tag{5.7}$$

The noise model adopted in this work has two components: (1) the background noise, which is assumed to be zero mean additive white Gaussian noise (AWGN) with variance σ_z^2, and (2) the impulsive noise, which is represented using the model in [22]. The narrowband interference is also considered as a cognitive system in the form of PU interference with variance σ_p^2.

5.4 Summary

In this chapter, a GSBPLM approach has been proposed, which represents the path loss of individual PLC channel classes. For this purpose, the corresponding nine classes of random topology generators have been built up, each class generator having an average attenuation profile similar to the corresponding PLC channel class. The GSBPLM is applicable to different scenarios and removes the contradiction between previous mapping approaches. The generated path loss curves for the BPLC are comparable to their counterparts for wireless communication in the VHF band, proving the feasibility of cooperation between the two communication technologies indoors. As well, the adopted VHF wireless complex indoor channel gain was presented together with the measurements of the crosstalk channel between the BPLC and TVWS channels. The measurements have shown convergence between the path loss of the cross-talk channel and the path loss of both the TVWS and BPLC channels. In Chapter 6, theoretical analysis will be presented for the cross talk between TVWS and BPLC channels through modeling the excited antenna current by the TVWS signal along the BPLC cables. The chapter will also investigate the conversion of this antenna mode current to differential mode, which is the dominant mode in the BPLC communications.

References

[1] Tlich, M., A. Zeddam, F. Moulin, and F. Gauthier, "Indoor Power-Line Communications Channel Characterization Up to 100 MHz Part I: One-Parameter Deterministic Model," *IEEE Transactions on Power Delivery*, Vol. 23, No. 3, July 2008, pp. 1392–1401.

[2] Zimmermann, M., and K. Dostert, "A Multipath Model for the Powerline Channel," *IEEE Transactions on Communications*, Vol. 50, No. 4, April 2002, pp. 553–559.

[3] Versolatto, F., and A. M. Tonello, "On the Relation between Geometrical Distance and Channel Statistics in In-Home PLC Networks," in *Proc. 16th IEEE International Symposium on Power Line Communications and Its Applications (ISPLC)*, March 2012, Beijing, China, pp. 280-285.

[4] Shin, J., J. Lee, and J. Jeong, "Channel Modeling for Indoor Broadband Power-Line Communications Networks with Arbitrary Topologies by Taking Adjacent Nodes Into Account," *IEEE Transactions on Power Delivery*, Vol. 26, No. 3, July 2011, pp. 1432–1439.

[5] Zheng, T., X. Yang, and B. Zhang, "Broadband Transmission Characteristics for Power-Line Channels," *IEEE Transactions on Power Delivery*, Vol. 21, No. 4, October 2006, pp. 1905–1911.

[6] Mustafa, H. D., S. Ashutosh, S. H. Karamchandani, S. N. Merchant, and U. B. Desai, "Non-Invasive, Reflection Coefficient Based Channel Estimation on PLC Systems," in *Proc. 8th International Conference on Information Communications and Signal Processing (ICICS)*, December 2011, Singapore, pp. 1–5.

[7] Tonello, A. M., and F. Versolatto, "Bottom-Up Statistical PLC Channel Modeling Part I: Random Topology Model and Efficient Transfer Function Computation," *IEEE Transactions on Power Delivery*, Vol. 26, No. 2, April 2011, pp. 891–898.

[8] Marrocco, G., D. Statovci, and S. Trautmann, "A PLC Broadband Channel Simulator for Indoor Communications," in *Proc. 17th IEEE International Symposium on Power Line*

Communications and Its Applications (ISPLC), March 2013, Johannesburg, South Africa, pp. 321–326.

[9] Andrusenko, J., R. L. Miller, J. A. Abrahamson, N. M. Merheb Emanuelli, R. S. Pattay, and R. M. Shuford, "VHF General Urban Path Loss Model for Short Range Ground-to-Ground Communications," *IEEE Transactions on Antennas and Propagation*, Vol. 56, No. 10, October 2008, pp. 3302–3310.

[10] Chen, S., "Ultra Wideband Gigabit Powerline Communication," Ph.D. Dissertation, Queen Mary University, London, UK, 2009.

[11] Lai, S. W., N. Shabehpour, G. G. Messier, and L. Lampe, "Performance of Wireless/Power Line Media Diversity in the Office Environment," in *Proc. IEEE Global Communications Conference (GLOBECOM)*, December 2014, Texas, pp. 2972–2976.

[12] Heggo, M., X. Zhu, Y. Huang, and S. Sun, "A Novel Statistical Approach of Path Loss Mapping for Indoor Broadband Power Line Communications," in *Proc. IEEE International Conference on Smart Grid Communications (SmartGridComm) 2014*, November 2014, Venice, Italy, pp. 499–504.

[13] Tao, Z., Y. Xiaoxian, and Z. Baohui, "Broadband Transmission Characteristics for Power-Line Channels," *IEEE Transactions on Power Delivery*, Vol. 21, No. 4, October 2006, pp. 1905–1911.

[14] Andrusenko, J., R. L. Miller, J. A. Abrahamson, N. M. Merheb Emanuelli, R. S. Pattay, and R. M. Shuford, "VHF general urban path loss model for short range ground-to-ground communications," *IEEE Transactions on Antennas and Propagation*, Vol. 56, No. 10, October 2008, pp. 3302–3310.

[15] Benvenuto N., and L. Tomba, "Performance Comparison of Space Diversity and Equalization Techniques for Indoor Radio Systems," *IEEE Transactions on Vehicular Technology*, Vol. 46, No. 2, May 1997, pp. 358–368.

[16] Karduri, M., M. D. Cox, and N. J. Champagne, "Near-Field Coupling between Broadband over Power Line (BPL) and High-Frequency Communication Systems," *IEEE Transactions on Power Delivery*, Vol. 21, No. 4, October 2006, pp. 1885–1891.

[17] Finamore,W. A., M. V. Ribeiro, and L. Lampe, "Advancing Power Line Communication: Cognitive, Cooperative, and MIMO Communication" *Proc. Brazilian Telecommunications Symposium*, September 2012, Brasilia, Brazil, pp. 13–16.

[18] Degauque, P., P. Laly, V. Degardin, M. Lienard, and L. Diquelou, "Compromising Electromagnetic Field Radiated by In-House PLC Lines," in *Proc. IEEE Global Telecommunications Conference (GLOBECOM)*, IEEE, December 2010, Florida, pp. 1–5.

[19] Degardin, V., P. Laly, M. Lienard, and P. Degauque, "Compromising Radiated Emission from a Power Line Communication Cable," *Journal of Communications Software & Systems*, Vol. 7, No. 1, March 2011, pp. 16–21.

[20] Degauque, P., P. Laly, V. Degardin, and M. Lienard, "Power line communication and compromising radiated emission," in *Proc. IEEE International Conference on Software, Telecommunications and Computer Networks (SoftCOM)*, IEEE, September 2010, Split-Bol, Croatia, pp. 88–91.

[21] Oliveira, T. R., C. A. G. Marques, M. S. Pereira, S. L. Netto, and M. V. Ribeiro, "The Characterization of Hybrid PLC-Wireless Channels: A Preliminary Analysis," in *Proc. 17th IEEE International Symposium on Power Line Communications and Its Applications (ISPLC)*, March 2013, Johannesburg South Africa. IEEE pp. 98– 102.

[22] Yin, J., X. Zhu, and Y. Huang, "Modeling of Amplitude-Correlated and Occurrence-Dependent Impulsive Noise for Power Line Communication," in *Proc. IEEE International Conference on Communications (ICC)*, June 2014, Sydney, Australia, pp. 4565–4570.

6

MIMO TV White Space–Broadband Power Line Communication Point-to-Point System

6.1 Introduction

In the previous chapter, both the BPLC and TVWS channels were studied thoroughly. A comparison was made in Chapter 5 between the path losses of both channels, and the convergence between them was shown. This proved the feasibility of having cooperative system between the TVWS and BPLC. Measurements carried out in Chapter 5 and the theoretical analysis presented had investigated the cross talk between both channels, which could be exploited in enhancing the received signal in an MIMO cooperative system. In this chapter, a new cooperative system is proposed between BPLC and TVWS in the VHF band 54 MHz–88 MHz referred to as white BPLC (WPLC). The proposed system is different from the previous MIMO cognitive BPLC system in [1] since the WBPLC Tx can access the VHF band with lower Tx PSD constraints using the TVWS standard [2], which significantly enhances the overall system capacity. First, to the best of our knowledge, this is the first work to propose a cognitive BPLC solution that complies with the TVWS standard. The addition of wireless antenna to cognitive BPLC is essential to enable the WBPLC Tx to sense the TV PU at very low power level of −114 dBm, which satisfies the requirement of the TVWS standard [2]. This addition of the TVWS antenna solves the problem of coupling loss that exists in previous cognitive BPLC solutions. Second, the interference to other wireless services is mitigated using a proposed iterative precoding technique at the Tx. The proposed algorithm takes the advantage of the MISO channel between the WBPLC Tx and the TV wireless Rx. Third, an efficient power allocation algorithm is derived for the MIMO system in order to achieve maximum ergodic capacity for each subcarrier. Fourth, compared to the hybrid BPLC WiFi system [3], the proposed system offers a more cost-effective hybrid solution. This is due to the elimination of the RF up- and downconverters needed for the WiFi transceiver.

The chapter is organized as follows. In Section 6.2, the proposed system model is presented. In Section 6.3, cognitive spectrum access is addressed, including an iterative precoding technique and MIMO spectrum sensing. In Section 6.4, the power allocation for the MIMO channel is presented. In Section 6.5, the capacity simulation results of the proposed system are compared with BPLC WiFi and MIMO BPLC. Finally, in Section 6.6 the chapter is summarized.

Notations: $\mathbb{E}\{.\}$ denotes the expectation operator, $\lceil x \rceil^+$ denotes $\max(0, x)$, vectors are represented by boldface letters and the Hermitian of a matrix $\mathbf{A}$ is $\mathbf{A}^\dagger$, and the conjugate transpose of a vector $\mathbf{v}$ is $\mathbf{v}^\dagger$. $\mathbf{I}$ denotes the identity matrix.

6.2 System Model

The proposed WBPLC system model is illustrated in Figure 6.1. The TV PU activity (i.e., presence or absence) in the TVWS band is assumed to be known at the WBPLC Tx due to access to a geolocation database as stated in [2]. In this chapter, two modes of operation are proposed for the TVWS BPLC system according to the available channel state information (CSI). The CSI of the PU Rx in this work is defined as the interference complex channel gain vector $\mathbf{g}$ between the WBPLC Tx and the PU Rx. The two proposed modes are: (1) opportunistic mode: The CSI of the PU Rx is available to the WBPLC transceiver and (2) nonopportunistic mode: The CSI of the PU Rx is unavailable to the WBPLC transceiver. The CSI in the opportunistic mode can be obtained using the sensing method in [4]. The method simply assumes that the SU joins the PU network in the first time slot, and does not send any signal. Let x_i

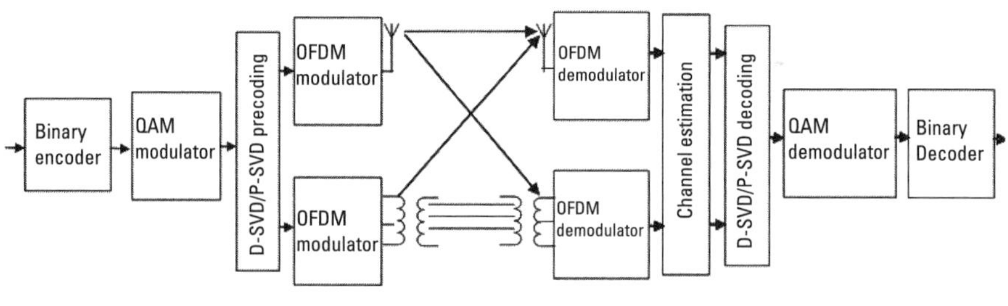

Figure 6.1 System model of WBPLC transceiver, using both BPLC and wireless TVWS channels.

and η_i represent the transmitted signal and the channel gain between the PU Tx and Rx at ith time slot, respectively. Hence, the received message by the PU Rx is:

$$y_i = \eta_i x_i + z_i \tag{6.1}$$

The PU Rx broadcasts a message containing the estimation of the channel gain $\hat{\eta}_i \approx \eta_i$, which is received by the SU Rx. In the second time slot, the PU message is received and decoded by the SU and forwarded again with power α. Let g represents the channel gain between the SU Tx and the PU Rx. Hence, the received message by the PU is

$$y_i = (\eta_{i+1} + g_{i+1}\alpha)x_{i+1} + z_{i+1} \tag{6.2}$$

The PU again broadcasts the estimation of the new channel gain $\hat{\eta}_{i+1} \approx \eta_{i+1} + g_{i+1}\alpha$, which is received by the SU. Now, the SU can estimate the channel gain g_{i+1} using $\hat{\eta}_i$ and $\hat{\eta}_{i+1}$. Obtaining the CSI between the WBPLC Tx and the PU Rx using the method in [4] relies on the permission of the PU network given to the SU in order to listen and add small interference noise power (α) to the PU communication. The advantage of this method is that it does not add any overhead communication to the PU (i.e., signaling between the PU and the SU). However, if this method is not allowed by the PU, then a switch to the nonopportunistic mode of operation can be done, which does not need obtaining the CSI between the WBPLC Tx and the PU Rx.

As illustrated in Figure 6.1, the binary encoded data is modulated using quadrature amplitude modulation (QAM) modulator. Each QAM symbol is precoded using either singular value decomposition (SVD) algorithm or projected singular value decomposition (P-SVD) algorithm. The precoding algorithm is discussed in detail in Section 6.3. The frame structure follows the IEEE 1901 [5], which contains two parts: the frame control part and the payload part. The former is sent over a separate control channel in the noncognitive frequency band below 30 MHz, while the latter is sent over the WBPLC cognitive channel with variable length according to the available time slots. The received signal is decoded using a decoding matrix according to the precoding type. The decoded symbols are passed through a maximum likelihood (ML) detector to allow estimation of the transmitted symbols.

Since the proposed system accesses the TVWS, a cognitive spectrum access algorithm is proposed. In the algorithm the channel between a single WBPLC Tx and Rx is considered to be $n_r \times n_r$ MIMO channel where n_r is the number of WBPLC MIMO subchannels as illustrated in Figure 6.2. The n_r MIMO subchannels are divided into two groups: (1) $(n_r - 1) \times (n_r - 1)$ MIMO BPLC sub-channels as in the case of 2×2 single phase MIMO BPLC and 3×3 three phase MIMO BPLC, and (2) single-input single-output (SISO) TVWS wireless

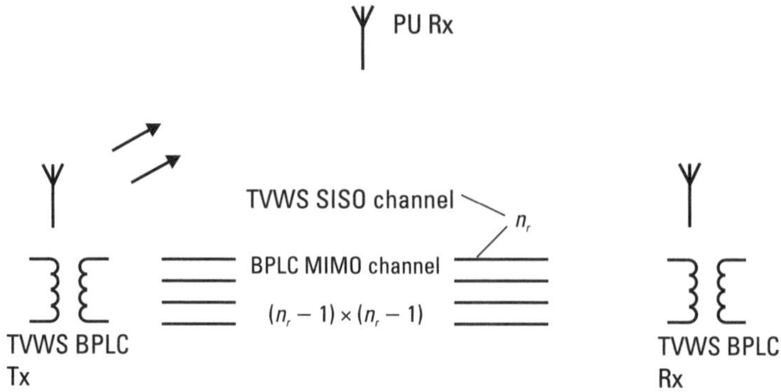

Figure 6.2 $n_r \times n_r$ WBPLC Tx-Rx channel model and $n_r \times 1$ WBPLC Tx-TV PU Rx channel model.

subchannel. However, the channel between a WBPLC Tx and a single TV PU Rx can be considered as $n_r \times 1$ MISO channel. This advantage is used by the proposed system for better WBPLC to TV PU interference mitigation. The signal received at the WBPLC Rx on subcarrier n can be represented as [6]

$$\mathbf{y}(n) = \tilde{\mathbf{H}}^{(n)}\mathbf{U}^{(n)}\mathbf{x}(n) + \mathbf{E}^{(n)}\mathbf{U}^{(n)}\mathbf{x}(n) + \mathbf{z}(n) \tag{6.3}$$

where $\tilde{\mathbf{H}}^{(n)}$ is the estimated $n_r \times n_r$ MIMO channel matrix for subcarrier n, which can be expressed as $\tilde{\mathbf{H}}^{(n)} = \mathbf{Q}^{(n)}(\Lambda^{(n)})^{1/2}\mathbf{U}^{\dagger(n)}$. $\mathbf{E}^{(n)} = \mathbf{Q}^{(n)}\tilde{\mathbf{E}}^{(n)}\mathbf{U}^{\dagger(n)}$ is a zero mean Gaussian distributed error matrix with the variance σ_E^2, which represents the error introduced in the channel estimation as previously addressed in [7–9], it is assumed that the channel transfer matrix $\mathbf{H}^{(n)} = \tilde{\mathbf{H}}^{(n)} + \mathbf{E}^{(n)}$. $\mathbf{U}^{(n)}$ is the precoding matrix for subcarrier n and $\mathbf{z}(n)$ is AGWN vector whose elements have zero mean and variance σ_z^2.

Without loss of generality, the antenna is assumed to be omnidirectional with unity gain [10], and the directional antenna can further improve the performance by introducing more antenna gain. In order to have more compact VHF antennas for practical applications, several researches had been done in [11] and [12] to design small VHF antennas (i.e., ≤0.65m. length). This can be exploited in the WBPLC system for more practical design and implementation.

Multiplying the received signal by the decoding matrix $\mathbf{Q}^{\dagger(n)}$. It can be deduced

$$\hat{\mathbf{y}}(n) = (\Lambda^{(n)})^{1/2}\mathbf{x}(n) + \tilde{\mathbf{E}}^{(n)}\mathbf{x}(n) + \hat{\mathbf{z}}(n) \tag{6.4}$$

where $\hat{\mathbf{z}}(n) = \mathbf{Q}^{(n)}\mathbf{z}(n)$ and $\Lambda^{(n)}$ is the matrix of the singular values of the MIMO channel power gain matrix $\mathbf{S}$ which can be defined as $\mathbf{S} = \tilde{\mathbf{H}}^{(n)}\tilde{\mathbf{H}}^{\dagger(n)}$.

6.3 Cognitive Spectrum Access

In this section, the cognitive spectrum access for the hybrid WBPLC Tx is presented, including (1) an iterative precoding technique, and (2) cognitive spectrum sensing.

6.3.1 Iterative Hybrid SVD/P-SVD Precoding Technique

In this work, a new precoding algorithm is proposed for cognitive spectrum access. The algorithm aims to achieve maximum capacity while maintaining the interference limit at the TV PUs Rxs by applying Lemma 1 that will be discussed later in this section. The algorithm precodes the WBPLC Tx data according to the WBPLC-TV PU channel. Two precoding algorithms are used: (1) SVD, and (2) P-SVD [13]. The SVD is used in the idle channel case while a hybrid SVD/P-SVD is used in case of an occupied channel. In case of SVD precoding, the channel matrix $\mathbf{H}^{(n)}$ is decomposed using SVD into $\mathbf{H}^{(n)} = \mathbf{Q}^{(n)}(\Lambda^{(n)})^{1/2}\mathbf{U}^{\dagger(n)}$ where $\mathbf{Q}^{(n)}$ and $\mathbf{U}^{(n)}$ are $n_r \times n_r$ matrices with orthonormal columns. Λ^n is an $n_r \times n_r$ diagonal positive matrix with vector $\lambda^{(n)}$ as its diagonal. Hence, $\mathbf{V} = \mathbf{U}^{(n)}$ will be the precoding matrix as shown in (6.3). However, in the P-SVD precoding, the SU MIMO channel is first projected on to the null space of the interference channel vector with the PU. Then, the projected channel matrix is then decomposed using the SVD method. Let $\hat{\mathbf{g}_1}^{(n)} = \mathbf{g}_1^{(n)}/\|\mathbf{g}_1^{(n)}\|$ be the unit vector in the direction of $\mathbf{g}_1^{(n)}$ that represents the channel vector between the WBPLC and the TV PUs. The subscript 1/0 in $g_{1/0}^{(n)}$ indicates the TV PU presence/absence, respectively. The projection of the channel matrix $\mathbf{H}^{(n)}$ can be defined in the null space of $\mathbf{g}_1^{\dagger(n)}$ as [13]

$$\mathbf{H}_{\perp}^{(n)} = \mathbf{H}^{(n)}(\mathbf{I} - \hat{\mathbf{g}_1}^{(n)}\hat{\mathbf{g}_1}^{\dagger(n)}) \tag{6.5}$$

The vector multiplication of $\hat{\mathbf{g}_1}^{(n)}\hat{\mathbf{g}_1}^{\dagger(n)}$ results in a matrix such that $(\hat{\mathbf{g}_1}^{(n)}\hat{\mathbf{g}_1}^{\dagger(n)})\hat{\mathbf{g}_1}^{(n)} = \hat{\mathbf{g}_1}^{(n)}$. Hence, $(\mathbf{I} - \hat{\mathbf{g}_1}^{(n)}\hat{\mathbf{g}_1}^{\dagger(n)})\hat{\mathbf{g}_1}^{(n)}$ is equal to zero. Consequently, $\mathbf{H}_{\perp}^{(n)}$ represents the projection of $\mathbf{H}^{(n)}$ matrix into the null space of $\hat{\mathbf{g}_1}^{\dagger(n)}$). Then using the SVD of $\mathbf{H}_{\perp}^{(n)} = \mathbf{Q}_{\perp}^{(n)}(\Lambda_{\perp}^{(n)})^{1/2}\mathbf{U}_{\perp}^{\dagger(n)}$ the precoding matrix can be concluded $\mathbf{V} = \mathbf{U}_{\perp}^{(n)}$ and decode the received signal by multiplying by $\mathbf{Q}_{\perp}^{\dagger(n)}$.

$$\hat{\mathbf{y}}(n) = (\Lambda_{\perp}^{(n)})^{1/2}\mathbf{x}(n) + \hat{\mathbf{z}}(n) \tag{6.6}$$

In the proposed precoding algorithm, both SVD and P-SVD are used jointly according to the CSI of both the WBPLC and TV PU Rxs.

6.3.1.1 Idle Channel

In case of an idle channel, SVD is used as a precoding scheme. Although the PU is absent, an interference limit is forced for false detection probability to avoid harmful interference. The WBPLC-TV PU channel $\mathbf{g}_{0,k}^{(n)}$ with maximum gain $\max_{n_t} \|\mathbf{g}_{0,k}^{(n)}\mathbf{U}_0^{(n)}\|^2$ can only be accessed with a power level below the interference limit where $k = 1, 2, ..n_t$ and n_t is the number of PUs.

6.3.1.2 Occupied Channel

This case is only applied to the opportunistic mode of operation mentioned in Section 6.2. In this case, the PU is present, while the secondary user is allowed also to access the channel such that the received power at the TV PU Rx is below a certain interference limit Γ. In the proposed algorithm the SVD is also used as a default precoding scheme as in the case of the idle channel. However, the decision is taken to switch the precoding scheme to P-SVD according to the CSI of the TV PU Rx. The decision is taken to satisfy two conditions: (1) achieve the interference limit at the TV PU Rx, and (2) achieve high signal to noise ratio (SNR) at the WBPLC Rx. Hence, the previous two conditions can be satisfied knowing the CSI as follows. Let's assume that the interference limit is the minimum threshold for the signal to be detected by the Rx. Hence, the interference to the PU should be below this threshold in order not to affect the PU reception, while the signal received at the WBPLC Rx should be above that threshold in order to be detected.

Lemma 1 Let γ be the SNR at the WBPLC Rx and Γ be the interference limit at the TV PU Rx. In order to achieve $\gamma \geq \Gamma$ while satisfying the interference limit at the TV PU Rx, the following condition should be satisfied

$$\det \begin{bmatrix} \frac{\lambda_1^{(n)}}{\rho} - \alpha_{1,k}^{(n)} & \cdots & -\alpha_{n_r,k}^{(n)} \\ \vdots & & \vdots \\ -\alpha_{1,k}^{(n)} & \cdots & \frac{\lambda_{n_r}^{(n)}}{\rho} - \alpha_{n_r,k}^{(n)} \end{bmatrix} \geq 0 \tag{6.7}$$

where the matrix diagonal is positive such that $\frac{\lambda^{(n)}}{\rho} \geq \alpha_{\mathrm{k}}{}^{(n)}$, ρ is a constant that satisfies $\rho \geq 1$, and $\alpha_k = \|\mathbf{g}_{1,k}^{(n)}\mathbf{U}_1^{(n)}\|^2$ of the kth PU. Lemma 1 implies that the WBPLC MIMO channel gain represented by its singular values $\lambda^{(n)}$ should be greater than gain of the interference channel with the PU represented by α_k. The ρ constant controls the gap between both channel gains.

Proof: See Appendix B.

Lemma 1 should be satisfied for the use of SVD. If this condition is violated, then P-SVD is used to eliminate the interference at one TV PU Rx. The eliminated PU is selected to satisfy $\max_{n_t}\{\|\alpha_{\mathrm{k}}{}^{(n)}\|^2\}$. Hence, the algorithm proposed for cognitive access can be summarized as

Proposed iterative hybrid SVD/P-SVD algorithm

- Initialize the precoding algorithm as SVD and $i = n_r$.
- While $i > 0$
  Check the condition in Lemma 1.
  - If the condition is satisfied, end the loop, otherwise change the precoding to P-SVD.
  - Select the WBPLC-TV PU interference channel to be canceled that satisfies $\max_{n_t}\{\|\alpha_{\mathrm{k}}{}^{(n)}\|\}$.
  - $i = i - 1$
- End.

6.3.2 Cognitive Spectrum Sensing

It is known that the sensing time is considered as a challenging issue for the cognitive systems. In the proposed system the optimum sensing time for MIMO cognitive system is achieved through achieving the optimum detection probability for each MIMO subchannel. This can be clarified in the following Lemma 2. Let $\hat{\mathcal{P}}_d$ represent the target detection probability of the MIMO cognitive Rx using the energy detection method. Also, let the OR rule (i.e., the channel is detected as occupied if the PU is detected using either the BPLC Rx or the TVWS Rx) be adopted for MIMO cognitive sensing. Hence, the relationship between the detection probability of each subchannel and the target detection probability of the MIMO channel can be expressed as

Lemma 2

$$\begin{aligned}\Pi_{i=1}^{n_r}\Big(1 - Q\Big[\frac{\gamma_m - \gamma_i}{\gamma_m\sqrt{2\gamma_i + 1}} Q^{-1}(\hat{\mathcal{P}}_{fa})\\ + Q^{-1}(\mathcal{P}_{d_m})\sqrt{\frac{2\gamma_m + 1}{2\gamma_i + 1}}\Big]\Big) = 1 - \hat{\mathcal{P}}_d\end{aligned} \quad (6.8)$$

where $\mathcal{P}_{d_m}$ is the detection probability of the mth MIMO subchannel, γ_m and γ_i are the SNR of the mth and ith subchannels, respectively for $i, m = 1, ..n_r$, $\hat{\mathcal{P}}_{fa}$ is the target false alarm of each MIMO subchannel and $Q(.)$ is the complementary error function.

Proof: See Appendix C based on energy detection method in [14].

Using Lemma 2, $\mathcal{P}_{d_m}$ can be derived numerically for any MIMO subchannel. Let $N_{min} = \tau_{min} f_s$ be the minimum number of samples requested to achieve the target detection probability $\hat{P}_d$, f_s is the sampling frequency and τ_{min} is the minimum sensing time. N_{min} can be expressed as [14]

$$N_{min} = \frac{1}{\gamma_m^2}\left(Q^{-1}(\mathcal{P}_{fa_m}) - Q^{-1}(\mathcal{P}_{d_m})\sqrt{2\gamma_m + 1}\right)^2 \tag{6.9}$$

N_{min} and τ_{min} can be directly evaluated after obtaining $\mathcal{P}_{d_m}$ numerically using Lemma 2.

6.4 Capacity Maximization Based Power Allocation

In this section, the capacity model and the power allocation algorithm are presented, which have been adopted to represent the MIMO WBPLC channel.

6.4.1 Capacity Analysis

In [15], a system model was proposed for the cognitive spectrum access, where the secondary user could access the cognitive band with two power levels $P_0^{(n)}$ and $P_1^{(n)}$ in the TV PU absence and presence, respectively. In this work, the model proposed in [15] is developed to represent the MIMO channel between WBPLC Tx and Rx. Also, the model is modified to represent the MISO interference channel between WBPLC and TV PU. Hence, the spectral efficiencies (SEs) on the n-th subcarrier for the four scenarios are given as

$$r_{c,d}^{(n)} = \frac{1}{N}\Sigma_{i=1}^{n_r} \log_2\left(1 + \frac{\lambda_{d,i}^{(n)} P_{d,i}^{(n)}}{\sigma_z^2 + \sigma_E^2 P_{av}^{(n)} + d\sigma_p^2}\right) \tag{6.10}$$

where c and d can take the values of 0 and 1 for the idle and occupied channel status, respectively. σ_z^2 and σ_p^2 are the noise power and the PU power, respectively. σ_E^2 is the variance of the channel estimation error $\mathbf{E}^{(n)}$ and $P_{av}^{(n)}$ is the maximum average power allowed for the WBPLC Tx for subcarrier n. $\lambda_{d,i}^{(n)}$ and $P_{d,i}^{(n)}$ are the singular value and the WBPLC Tx signal power of the ith MIMO sub-channel and nth subcarrier assigned to each detected channel status d.

Hence, the MIMO ergodic SE can be concluded as

$$R^{(n)} = \frac{\Pi_{I0}}{N}\left(\frac{T-\tau}{T}\right)\mathbb{E}_{\mathbf{H}^{(n)},\mathbf{g}^{(n)}}\Big\{\mathcal{P}(H_0)(1-\mathcal{P}_{fa})r_{00}^{(n)} + \mathcal{P}(H_0)\mathcal{P}_{fa}r_{01}^{(n)} + \mathcal{P}(H_1)(1-\mathcal{P}_d)r_{10}^{(n)} + \mathcal{P}(H_1)\mathcal{P}_d r_{11}^{(n)}\Big\} \tag{6.11}$$

Hence, the overall capacity in (bits per seconds) for all the subcarriers can be expressed as

$$C = \frac{B\Pi_{I0}}{N}\left(\frac{T-\tau}{T}\right)\Sigma_{n=1}^{N}\mathbb{E}_{\mathbf{H}^{(n)},\mathbf{g}^{(n)}}\Big\{\mathcal{P}(H_0)(1-\mathcal{P}_{fa})r_{00}^{(n)} + \mathcal{P}(\mathcal{H}_0)\mathcal{P}_{fa}r_{01}^{(n)} + \mathcal{P}(\mathcal{H}_1)(1-\mathcal{P}_d)r_{10}^{(n)} + \mathcal{P}(\mathcal{H}_1)\mathcal{P}_d r_{11}^{(n)}\Big\} \quad (6.12)$$

where $\mathcal{P}(\mathcal{H}_0)$ and $\mathcal{P}(\mathcal{H}_1)$ are the probabilities of TV PU absence and presence, respectively. $\mathcal{P}_d$ and $\mathcal{P}_{fa}$ are the TV PU detection probability and false alarm probability, respectively. T is the symbol time duration and τ is the TV PU detection time duration. Π_{I0} is the steady-state probability of the nonburst impulsive state $I0$ of the BPLC channel [16]. It is worth mentioning that due to low occurrence probability of impulsive bursts, Π_{I0} can take values that are close to one [16]. B is the total channel bandwidth of the OFDM symbol and N is the total number of subcarriers. For a constrained power communication system, the average MIMO power of the WBPLC per subcarrier should be less than a predefined value P_{av} as

$$\mathbb{E}_{\mathbf{H}^{(n)},\mathbf{g}^{(n)}}\{\mathcal{P}(\mathcal{H}_0)(1-\mathcal{P}_{fa})\Sigma_{i=1}^{n_r}P_{0,i}^{(n)} + \mathcal{P}(\mathcal{H}_0)\mathcal{P}_{fa}\Sigma_{i=1}^{n_r}P_{1,i}^{(n)} + \mathcal{P}(\mathcal{H}_1)(1-\mathcal{P}_d)\Sigma_{i=1}^{n_r}P_{0,i}^{(n)} + \mathcal{P}(\mathcal{H}_1)\mathcal{P}_d\Sigma_{i=1}^{n_r}P_{1,i}^{(n)}\} \leq P_{av} \quad (6.13)$$

The MISO interference of each subcarrier to the TV PU is also limited to a certain value Γ as

$$\mathbb{E}_{\mathbf{H}^{(n)},\mathbf{g}^{(n)}}\{\mathcal{P}(\mathcal{H}_1)(1-\mathcal{P}_d)\|\mathbf{g}_0^{(n)}\mathbf{U}_0^{(n)}\|^2\mathbf{P}_0^{(n)} + \mathcal{P}(\mathcal{H}_1)\mathcal{P}_d\|\mathbf{g}_1^{(n)}\mathbf{U}_1^{(n)}\|^2\mathbf{P}_1^{(n)}\} \leq \Gamma \quad (6.14)$$

$\mathbf{U}_0^{(n)}$ and $\mathbf{U}_1^{(n)}$ are the precoding matrices in case of TV PU absence and TV PU presence, respectively. Also $\mathbf{g}_0^{(n)}$ and $\mathbf{g}_1^{(n)}$ are the selected WBPLC-TV PU channel gain for the cases of TV PU absence and presence, respectively. The criterion of selection is according to the precoding scheme as mentioned in Section 6.3. $\mathbf{P}_0^{(n)}$ and $\mathbf{P}_1^{(n)}$ are the WBPLC signal power vectors in case of TV PU absence and presence, respectively, where $P_{0,i}^{(n)}$ and $P_{1,i}^{(n)}$ are their ith elements, respectively.

6.4.2 Power Allocation

In [15] a power allocation algorithm was proposed for SISO cognitive system. In this work the allocated power is derived for each cognitive radio MIMO subchannel (TVWS and BPLC) taking into consideration the use of SVD/P-SVD as a precoding technique. The power is allocated to each subchannel in both cases of TV PU presence and absence. The power allocation is done in a way to maximize the WBPLC MIMO capacity for each subcarrier and satisfy both WBPLC average power and TV PU Rx interference power limit. Hence, the problem of allocating the power can be formulated as

$$\max_{P_{0,i}^{(n)},P_{1,i}^{(n)}} (R^{(n)}) = \left(\frac{T-\tau}{T}\right)\mathbb{E}_{\mathbf{H}^{(n)},\mathbf{g}^{(n)}}\{\mathcal{P}(\mathcal{H}_0)(1-\mathcal{P}_{fa})r_{00}^{(n)} + \mathcal{P}(\mathcal{H}_0)\mathcal{P}_{fa}r_{01}^{(n)} + \mathcal{P}(\mathcal{H}_1)(1-\mathcal{P}_d)r_{10}^{(n)} + \mathcal{P}(\mathcal{H}_1)\mathcal{P}_d r_{11}^{(n)}\} \tag{6.15}$$

Subject to (6.13), (6.14), and $P_{0,i}^{(n)}, P_{1,i}^{(n)} \geq 0$, where the $\frac{\Pi_{I0}}{N}$ is omitted to simplify the derivation. The dual Lagrangian function can be represented as

$$\begin{aligned}
L(P_{0,i}^{(n)}, P_{1,i}^{(n)}, u, v) = {} & \left(\frac{T-\tau}{T}\right)\mathbb{E}_{\mathbf{H}^{(n)},\mathbf{g}^{(n)}}\{\mathcal{P}(\mathcal{H}_0)(1-\mathcal{P}_{fa})r_{00}^{(n)} + \mathcal{P}(\mathcal{H}_0)\mathcal{P}_{fa}r_{01}^{(n)} \\
& + \mathcal{P}(\mathcal{H}_1)(1-\mathcal{P}_d)r_{10}^{(n)} + \mathcal{P}(\mathcal{H}_1)\mathcal{P}_d r_{11}^{(n)}\} - v\mathbb{E}_{\mathbf{H}^{(n)},\mathbf{g}^{(n)}}\{ \\
& \mathcal{P}(\mathcal{H}_0)(1-\mathcal{P}_{fa})\Sigma_{i=1}^{n_r}P_{0,i}^{(n)} + \mathcal{P}(\mathcal{H}_0)\mathcal{P}_{fa}\Sigma_{i=1}^{n_r}P_{1,i}^{(n)} \\
& + \mathcal{P}(\mathcal{H}_1)(1-\mathcal{P}_d)\Sigma_{i=1}^{n_r}P_{0,i}^{(n)} + \mathcal{P}(\mathcal{H}_1)\mathcal{P}_d\Sigma_{i=1}^{n_r}P_{1,i}^{(n)}\} \\
& + vP_{av} - u\mathbb{E}_{\mathbf{H}^{(n)},\mathbf{g}^{(n)}}\{\mathcal{P}(\mathcal{H}_1)(1-\mathcal{P}_d)\|\mathbf{g}_0^{(n)}\mathbf{U}_0^{(n)}\|^2\mathbf{P}_0^{(n)} \\
& + \mathcal{P}(\mathcal{H}_1)\mathcal{P}_d\|\mathbf{g}_1^{(n)}\mathbf{U}_1^{(n)}\|^2\mathbf{P}_1^{(n)}\} + u\Gamma
\end{aligned} \tag{6.16}$$

The dual problem can be represented as

$$d(v,u) = \sup_{P_{0,i}^{(n)},P_{1,i}^{(n)}} L(P_{0,i}^{(n)}, P_{1,i}^{(n)}, u, v) \tag{6.17}$$

In order to obtain the supremum of the Langrangian with respect to the transmission powers, the primal dual decomposition can be used. Let $i \in 1, 2, .., n_r$ be an index for the MIMO subchannel and $j \in 0, 1$ be an index for the TV PU status (i.e., 0 in case of absence and 1 in case of presence). Also, let $\eta_0 = \mathcal{P}(\mathcal{H}_0)(1-\mathcal{P}_{fa})$, $\eta_1 = \mathcal{P}(\mathcal{H}_0)\mathcal{P}_{fa}$, $\beta_0 = \mathcal{P}(\mathcal{H}_1)(1-\mathcal{P}_d)$, and $\beta_1 = \mathcal{P}(\mathcal{H}_1)\mathcal{P}_d$. Using the primal dual decomposition, the joint variable optimisation problem can be subdivided into $n_r \times 2$ single variable convex optimization problems as

$$\max_{P_{j,i}^{(n)}\geq 0} (f(P_{j,i}^{(n)})) = \left(\frac{T-\tau}{T}\right)\mathbb{E}_{\mathbf{H}^{(n)},\mathbf{g}^{(n)}}\{\eta_j r_{0j,i}^{(n)} + \beta_j r_{1j,i}^{(n)}\} - v\mathbb{E}_{\mathbf{H}^{(n)},\mathbf{g}^{(n)}}\{(\eta_j+\beta_j)P_{j,i}^{(n)}\} - u\mathbb{E}_{\mathbf{H}^{(n)},\mathbf{g}^{(n)}}\{\beta_j\alpha_{j,i,k}^{(n)}P_{j,i}^{(n)}\} \tag{6.18}$$

Using the subgradient method, the solution to each problem can be addressed as

$$P_{j,i}^{(n)} = \lceil A_{i,j} + \sqrt{B_{i,j}}\rceil^{+} \tag{6.19}$$

where

$$A_{i,j} = \frac{\left(\frac{T-\tau}{T}\right)\log_2(e)(\eta_j + \beta_j)}{v(\eta_j + \beta_j) + u\beta_j\alpha_{j,i,k}^{(n)}} + \frac{2(\sigma_z^2 + \sigma_E^2 P_{av}^{(n)}) + \sigma_p^2}{\lambda_{j,i}^{(n)}} \tag{6.20}$$

$$B_{i,j} = A_{i,j}^2 - \frac{4}{\lambda_{j,i}^{(n)}}\Bigg[\frac{(\sigma_z^2 + \sigma_E^2 P_{av}^{(n)}) + \sigma_p^2}{\lambda_{j,i}^{(n)}\sigma_p^2} - \frac{\left(\frac{T-\tau}{T}\right)\log_2(e)(\eta_j((\sigma_z^2 + \sigma_E^2 P_{av}^{(n)}) + \sigma_p^2) + \beta_j(\sigma_z^2 + \sigma_E^2 P_{av}^{(n)}))}{v(\eta_j + \beta_j) + u\beta_j\alpha_{j,i,k}^{(n)}}\Bigg] \tag{6.21}$$

6.5 Simulation Results

6.5.1 Simulation Setup

In this section, the simulation results are presented for the proposed system compared to the MIMO BPLC [1] and the BPLC WiFi [3]. The OFDM symbol duration T is taken to be 5 ms. It is assumed that the BPLC channel as well as its noise has cyclostationary behavior with a coherence time that is typically half of the mains period (i.e., 10 ms) [17]. For the VHF wireless channel, a Rayleigh fading channel is assumed with average path loss as in (2.1), the l, α and d_0 are set to 1.5, 0.65, and 12m, respectively [18]. The total bandwidth from 54 MHz to 88 MHz is divided into five channels. Each channel has 6-MHz bandwidth. The transmission PSDs for the WBPLC, MIMO BPLC, and BPLC WiFi are set to −47 dBm/Hz, −85 dBm/Hz, and −50 dBm/Hz, respectively. The received SNR (i.e., the ratio of the received signal power to the power sum of the AWGN and the channel estimation error variance, regardless of the PU interference power) varies from 20 dB to 40 dB. The requested probability of false alarm and detection probability are 10^{-7} and 0.9999, respectively. As well, the probability of TV PU presence is 0.4. The received PU SNR is assumed to be −14 dB. n_r and n_t are assumed to equal 2. As well, energy detection algorithm is adopted for cognitive sensing and the OR-rule technique is adopted for MIMO sensing.

For the simulation of the power line environment, the random class topology generator presented in Chapter 5 is used. At each iteration, a random power line topology is generated and four points are randomly selected to represent WBPLC Tx and Rx and two TV PU Rxs.

6.5.2 Simulation Results

6.5.2.1 Capacity and Spectral Efficiency at Different Frequency Band Coverage

In Figure 6.3, the capacity complementary cumulative distribution function (CCDF) of the MIMO BPLC [1] and the BPLC WiFi [3] is compared for different

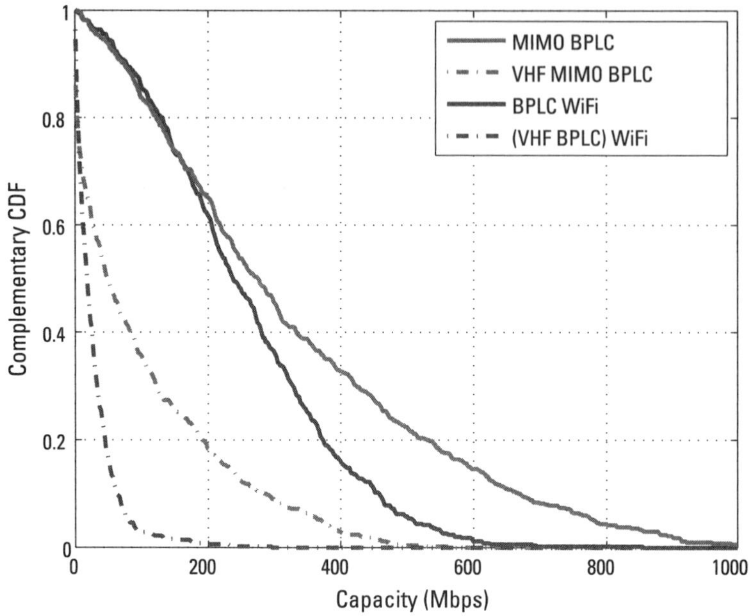

Figure 6.3 Complementary CDF versus ergodic capacity at different frequency bands.

frequency bands. The degradation in the performance is observed in the case of using VHF BPLC compared to the conventional BPLC. In the conventional BPLC, both the HF band below 30 MHz and the VHF band are allowed. Hence, more transmission power can be allocated to the HF band of the conventional BPLC leading to more capacity compared to the VHF BPLC.

6.5.2.2 WBPLC Spectral Efficiency

In Figure 6.4, the measured channel gain of the cross-talk channel between BPLC and wireless TVWS channel is shown for variable distances in the VHF band based on our previous measurements that were discussed in Chapter 5. In Figure 6.5, the average SEs of the SISO BPLC and wireless TVWS channels are compared with their counterparts in case of the cross talk channels. The SE of the full MIMO WBPLC system is shown in Figure 6.5. The comparison clarifies the benefit of combining both the TVWS and BPLC channels by compromising the radiation accompanying the BPLC signal and further exploiting the cross talk channels.

In Figure 6.6, the average SE of the proposed system is compared to the MIMO BPLC [1] and the BPLC WiFi [3] in the VHF band. Two notable observations can be addressed: First, the proposed system significantly improves the SE compared to the MIMO BPLC and the BPLC WiFi. This is due to compliance with the TVWS standard, which allows a transmission power of

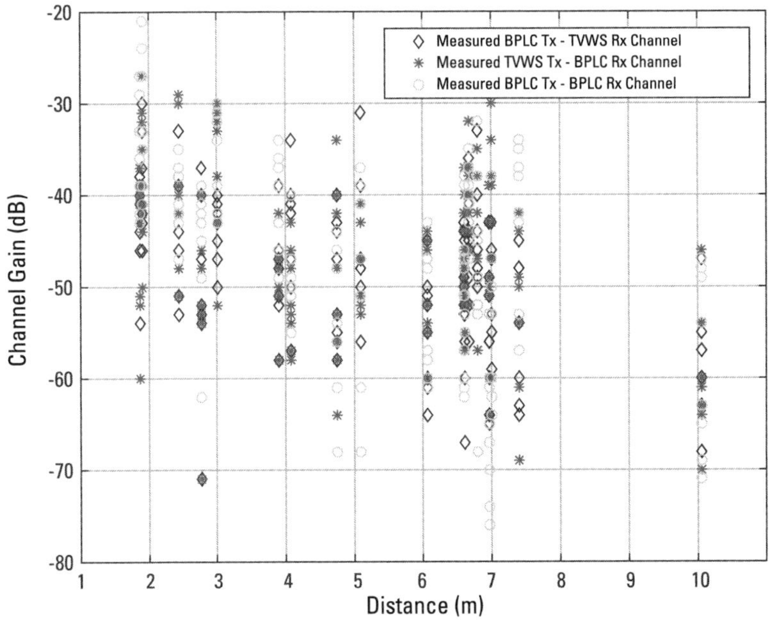

Figure 6.4 Measured channel gain of the cross-talk channel between BPLC and wireless TVWS compared to BPLC channel gain.

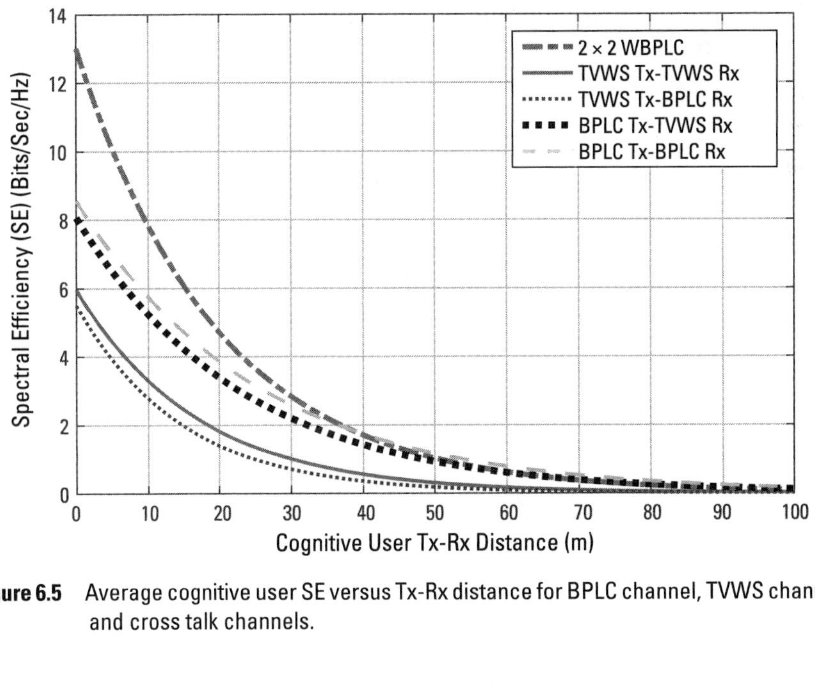

Figure 6.5 Average cognitive user SE versus Tx-Rx distance for BPLC channel, TVWS channel, and cross talk channels.

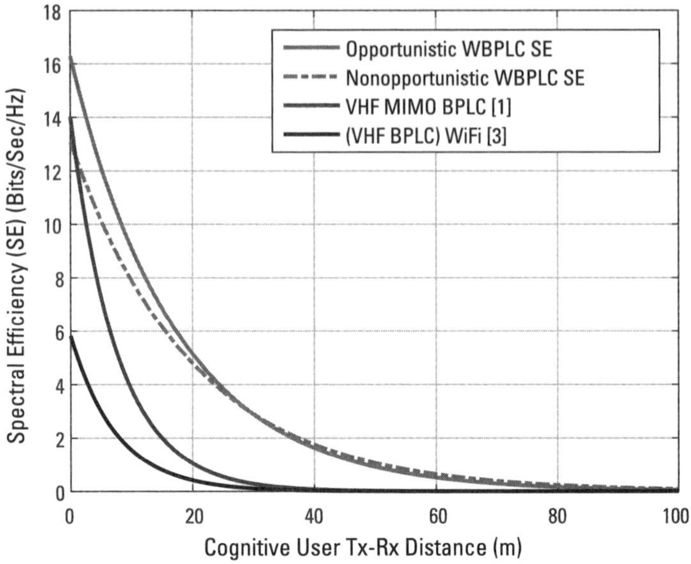

Figure 6.6 Average cognitive user SE versus Tx-Rx distance.

100 mW per 6-MHz channel and yet a PSD of −47 dBm/Hz. This yields a 38-dB increase in the PSD in the VHF band compared to the IEEE 1901 standard used in the MIMO BPLC. The VHF band has better channel gain than the 2.45-GHz band used in the BPLC WiFi. Second, a slight degradation in the performance in the WBPLC for the nonopportunistic case compared to the opportunistic case. This is due to the PU user interference in the opportunistic case that decreases the achievable capacity in the presence of the PU.

In Figure 6.7, the proposed WBPLC SE shows a rapid saturation to the maximum as the distance ratio increases. This proves that the proposed system including the two modes of operation are less interfering to the PU Rx even if they are located at very close distance to the PU Rx. This is due to the use of the proposed iterative precoding algorithm that has the ability to mitigate the interference with the nearby TV PU Rx.

6.5.2.3 Coupling Loss versus Sensing Time

The coupling loss represents the power loss due to coupling between the BPLC channel and the VHF wireless channel. As the coupling increases the coupling loss decreases and vice versa. In Figure 6.8, the effect of increasing the coupling loss on increasing the sensing time can be shown for the conventional cognitive BPLC. However, for the proposed system since MIMO sensing scheme is used with the aid of the TVWS antenna, the increase in the coupling loss does not affect the overall sensing time.

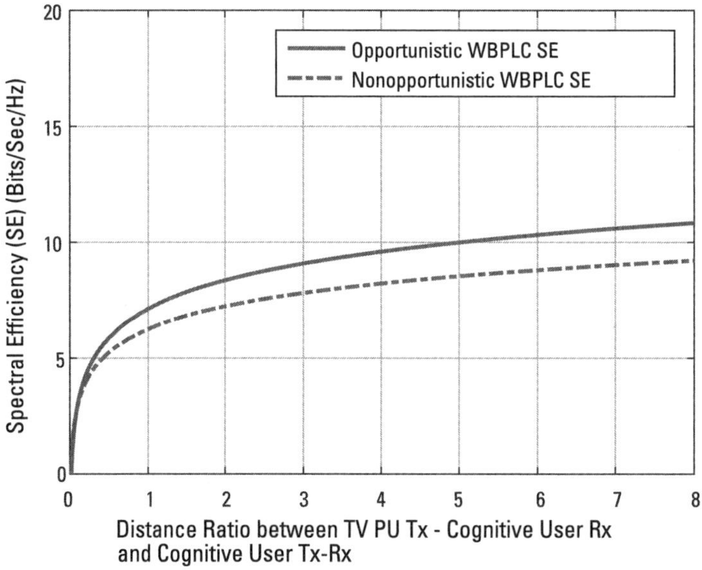

Figure 6.7 Average cognitive user SE versus the ratio between the TV PU Rx cognitive user Tx distance and the cognitive user Tx-Rx distance.

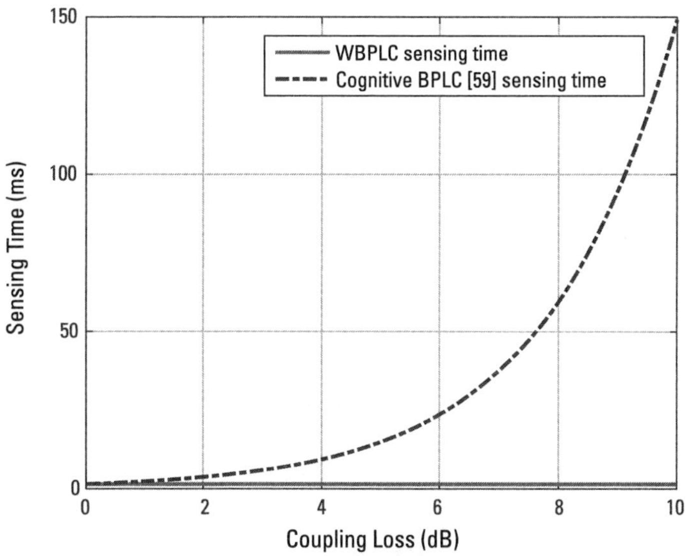

Figure 6.8 Spectrum sensing time versus coupling loss.

6.5.2.4 Spectral Efficiency versus Channel Estimation Error

In Figure 6.9, the effect of the channel estimation error is shown. The simulated values of σ_E^2 represent the measured values of channel estimation error variance

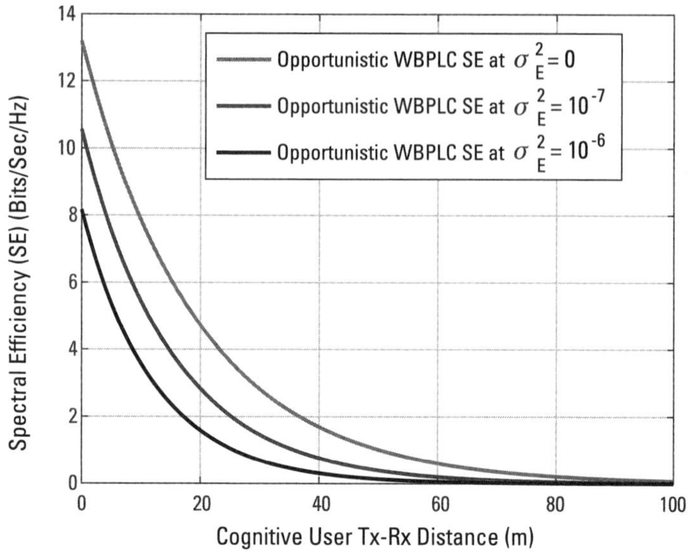

Figure 6.9 Average cognitive user SE versus Tx-Rx distance at different CSI error values.

for the indoor environment in the VHF band [9]. The ergodic capacity of both the WBPLC and the cognitive BPLC is degraded to 0.6 of its value on increasing the σ_E^2 from 0 to 10^{-6}. However, the WBPLC still outperforms the conventional cognitive BPLC even at high channel estimation error.

6.5.2.5 Ergodic Capacity of the VHF Band

In this part, the ergodic capacity is computed for the VHF band between 54 MHz and 88 MHz.

In Figure 6.10, the CCDF is presented versus the capacity. It can be observed that for the conventional cognitive BPLC system less than 0.2% of the users can achieve 200 Mbps, for the BPLC WiFi 10% can achieve 200 Mbps, and for the MIMO BPLC 30% can achieve the same capacity. However, for the proposed system more than 95% of the users can achieve 200 Mbps and more than 50% can achieve 400 Mbps.

6.5.3 Discussion on Practical Implementation and Measurement Results

The enhancement in the ergodic capacity achieved by the WBPLC transceiver is for two main reasons: (1) the compliance of the system with the TVWS standard [2], which allows higher Tx PSD compared to IEEE 1901 [5] or HomePlug AV2 [1], and (2) the enhancement in the sensing capability by using a wireless antenna rather than using the BPLC coupling circuit. Hence, the coupling loss introduced by both power line cables and BPLC coupling circuit

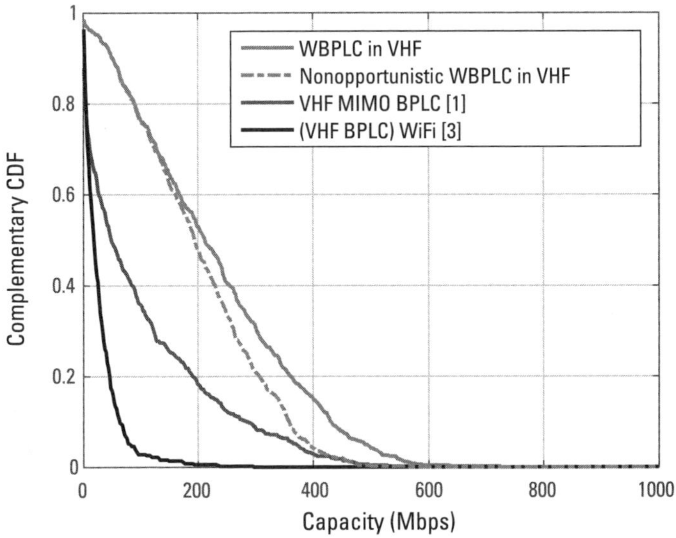

Figure 6.10 Complementary CDF versus ergodic capacity in the VHF band.

can be avoided. However, in certain cases, some TV frequency channels are not allowed to be accessed in a cognitive mode under the TVWS standard [2]. For example, fixed cognitive devices are not allowed to access the TVWS channels in the VHF that is adjacent to TV channels in operation [2]. In such cases, the WBPLC Tx can send data using the IEEE 1901 standard with a Tx PSD of -85 dBm/Hz, which can sufficiently protect the TV PU Rx. Since OFDM is adopted in the WBPLC transceiver, the PSD of the Tx can be controlled according to the allowed TV frequency channels that can be accessed under a cognitive regime.

6.6 Summary

In this chapter, a novel hybrid WBPLC point-to-point system was proposed for the indoor communication network. The proposed system offers a cost-effective solution for enhancing the BPLC capacity in the VHF band. An iterative precoding algorithm was also proposed to mitigate the interference with the TV PU Rxs in the VHF band. Moreover, an MIMO power allocation algorithm has been developed for the cognitive radio system. Through simulations it has been shown that the proposed system improves the BPLC capacity significantly. In the simulations, a single BPLC subchannel has been considered to form a 2×2 WBPLC system. The capacity of the proposed system can be further enhanced by using MIMO BPLC together with the TVWS wireless channel. The capacity enhancement has been preserved under small separation distances from the TV PU Rx. Moreover, the proposed system demonstrates robust performance against the variation in the BPLC coupling loss that affects the cognitive BPLC sensing time and capacity in the VHF band. In Chapter 7, this system will be developed to include other parameters such as point-to-multipoint and throughput maximization for a given bit error rate.

References

[1] Berger, L. T., A. Schwager, P. Pagani, and D.Schneider, *MIMO Power Line Communications: Narrow and Broadband Standards, EMC, and Advanced Processing*, Boca Raton, FL: CRC Press, 2014.

[2] Federal Communication Commission, Second Report and Order and Memorandum Opinion and Order in the Matter of Unlicensed Operation in the TV Broadcast Bands, Additional Spectrum for Unlicensed Devices Below 900 MHz and in the 3 GHz Band, Document 08-260, November 2008.

[3] Lai, S. W., N. Shabehpour, G. G. Messier, and L. Lampe "Performance of Wireless/Power Line Media Diversity in the Office Environment," in *Proc. IEEE Global Communications Conference (GLOBECOM)*, December 2014, Texas, pp. 2972–2976.

[4] Jovicic, A., and P. Viswanath, "Cognitive Radio: An Informationtheoretic Perspective," *IEEE Transactions on Information Theory*, Vol. 55, No. 9, September 2009, pp. 3945–3958.

[5] IEEE Standards Association et al. IEEE Standard for Broadband over Power Line Networks: Medium Access Control and Physical Layer Specifications, IEEE Std 1901, 2010, 2010, pp. 1–1586.

[6] Yoo, T., and A. Goldsmith, "Capacity and Power Allocation for Fading MIMO Channels with Channel Estimation Error," *IEEE Transactions on Information Theory*, Vol. 52, No. 5, May 2006, pp. 2203–2214.

[7] Grant, S. J., and J. K. Cavers, "Performance Enhancement through Joint Detection of Cochannel Signals Using Diversity Arrays," *IEEE Transactions on Communications*, Vol. 46, No. 8, August 1998, pp. 1038–1049.

[8] Zhu, X., and R. D. Murch, "Performance Analysis Of Maximum Likelihood Detection in a MIMO Antenna System," *IEEE Transactions on Communications*, Vol. 50, No. 2, February 2002, pp. 187–191.

[9] Etkin, R. H., and D. N. C. Tse, "Degrees of Freedom in Some Underspread MIMO Fading Channels," *IEEE Transactions on Information Theory*, Vol. 52, No. 4, April 2006, pp. 1576–1608.

[10] Davies, J., *Newnes Radio Engineer's Pocket Book*, Oxford, United Kingdom: Elsevier, 2014.

[11] Wongsakulphasatch, P., C. Phongcharoenpanich, and S. Kawdungta, "Compact Flat Monopole Antenna for Small Aircraft of VHF Communication System," in *Proc. IEEE Asia-Pacific Conference on Antennas and Propagation (APCAP)*, August 2012, Kaohsiung, Taiwan, pp. 27–28.

[12] Ryu, H.-K., G. Jung, and J.-M. Woo, "A Small Quarter Wavelength Microstrip Antenna for HF and VHF Band Applications," in *Proc. 10th Mediterranean Microwave Symposium*, August 2010, Guzelyurt, Turkey, pp. 48–51.

[13] Zhang, R., and Y.-C. Liang, "Exploiting Multi-Antennas for Opportunistic Spectrum Sharing in Cognitive Radio Networks," *IEEE Journal of Selected Topics in Signal Processing*, Vol. 2, No. 1, February 2008, pp. 88–102.

[14] Liang, Y.-C., Y. Zeng, E. C. Y. Peh, and A. T. Hoang, "Sensing-Throughput Tradeoff for Cognitive Radio Networks," *IEEE Transactions on Wireless Communications*, Vol. 7, No. 4, April 2008, pp. 1326–1337.

[15] Stotas S., and A. Nallanathan, "Enhancing the Capacity of Spectrum Sharing Cognitive Radio Networks," *IEEE Transactions on Vehicular Technology*, Vol. 60, No. 8, October 2011, pp. 3768–3779.

[16] Yin, J., X. Zhu, and Y. Huang, "Modeling of Amplitude-Correlated and Occurrence-Dependent Impulsive Noise for Power Line Communication," in *Proc. IEEE International Conference on Communications (ICC)*, June 2014, Sydney, Australia, pp. 4565–4570.

[17] Ferreira, H. C., L. Lampe, J. Newbury, and T. Swart (eds.), *Power Line Communications: Theory and Applications for Narrowband and Broadband Communications over Power Lines*, Chichester, United Kingdom: John Wiley & Sons, 2010.

[18] Andrusenko, J., R. L. Miller, J. A. Abrahamson, N. M. Merheb Emanuelli, R. S. Pattay, and R. M. Shuford, "VHF General Urban Path Loss Model for Short Range Ground-to-Ground Communications," *IEEE Transactions on Antennas and Propagation*, Vol. 56, No. 10, October 2008, pp. 3302–3310.

7

TV White Space Regulated Broadband Power Line Communication for IoT Networks: A Standard Perspective

7.1 Introduction

In this chapter, to mitigate the interference challenge in realizing an IoT network, we propose a standard framework that integrates the HomeplugAV2 standard [1,2] for BPLC and the ECMA-392 standard [3] for TVWS into a point-to-multipoint communication system, which is referred to as high-throughput white BPLC (HT-WBPLC). This work is different from the previous work in the following aspects:

1. To the best of our knowledge, this is the first work to investigate the integration of BPLC with TVWS in a downlink point-to-multipoint communication system, which integrates the requirement of primary user sensing and the permissible transmission PSD for TVWS users into the BPLC standard. Based on standard compatibility and our channel measurement results, the proposed standard framework enables additional frequency band of 100 MHz–200 MHz to be used by BPLC, guarantees minimum interference level between TVWS and BPLC, and allows higher transmission PSD for BPLC users in VHF. Both coverage and throughput analysis are provided for the proposed HT-WBPLC system. The results obtained show the benefit of utilizing the cross talk between wireless TVWS and BPLC to enhance the overall HT-WBPLC system performance, especially in the frequency beyond 100 MHz.

HT-WBPLC achieves higher throughput and coverage compared to MIMO BPLC.

2. The proposed HT-WBPLC standard framework allows three point-to-multipoint operating modes: (1) BPLC in HF, (2) TVWS in VHF and in PU absence, and c) BPLC in VHF and in PU presence, according to the PU status and operating frequency band, leading to a higher degree of freedom in adapting to different standard requirements of BPLC and TVWS. Resource allocation for the proposed HT-WBPLC standard framework is investigated to maximize the system throughput under different requirements of operating modes, while the previous work on resource allocation for orthogonal frequency division multiple access (OFDMA) based cognitive radio networks [4–9] cannot be applied directly, as each operating mode has their specific PSD and subchannel frequency spacing.

The rest of the chapter is organized as follows. In Section 7.2, we present the proposed HT-WBPLC system model for IoT network. In Section 7.3, channel measurement results for the cross talk between wireless TVWS channel and BPLC channel are presented. In Section 7.4, frequency and power allocation problem for different users is investigated and optimal solution is proposed. In Section 7.5, point-to-multipoint system simulation results are presented to evaluate the performance of our proposed HT-WBPLC system and in comparison to TVWS and MIMO BPLC systems. In Section 7.6, the chapter is summarized.

7.2 HT-WBPLC: Standard Overview and System Model

7.2.1 HT-WBPLC Standard for IoT Networks

The first edition of the ECMA-392 [3] standard was released in 2009 for TVWS communications, and the second edition was released in 2012. HomplugAV2 standard [2] was issued in 2012 as an extension to the IEEE 1901 standard for BPLC systems. The common band between ECMA-392 and HomeplugAV2 standards is the VHF band. However, BPLC and TVWS transceivers are permitted to access the VHF band at different transmission PSD levels. A BPLC transceiver uses −80 dBm/Hz PSD to access VHF band compared to −47.7 dBm/Hz PSD, in the case of TVWS transceiver.

The proposed standard amendment for HT-WBPLC allows the integration of TVWS and BPLC into one communication system in IoT network. Hence, it incorporates both ECMA-392 and HomeplugAV2 standards. HT-WBPLC has three modes of operation according to the frequency band and PU activity. In Figure 7.1, different operating modes of HT-WBPLC are mapped over four quadrants, where the upper part of the figure represents PU absence case and the

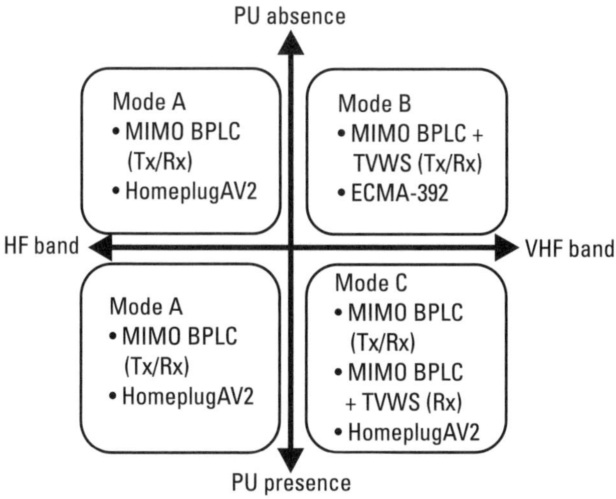

Figure 7.1 HT-WBPLC modes of operation.

right part represents VHF band access and vice versa. The three HT-WBPLC modes are explained below.

7.2.1.1 Mode A

Mode A spans the frequency band 1.8 MHz–30 MHz. Consequently, the HomplugAV2 standard is adopted for the PHY and MAC layers in communication with a transmission signal PSD of −55 dBm/Hz. The band below 30 MHz has the advantage that it is a completely free band, which can provide a backup channel when all TVWS channels are occupied. In WBPLC, SISO BPLC is used for communication, while in the HT-WBPLC, MIMO BPLC is used.

7.2.1.2 Mode B

Mode B deals with the case of PU absence in the VHF band. Therefore, the spanned frequency spectrum is 54 MHz–200 MHz. According to FCC regulations, nine TV channels are allowed for cognitive access in this band, which are 54 MHz–60 MHz, 60 MHz–66 MHz, 66 MHz–72 MHz, 76 MHz–82 MHz, 82 MHz–88 MHz, 174 MHz–180 MHz, 180 MHz–186 MHz, 186 MHz–192 MHz and 192 MHz–198 MHz. The ECMA-392 standard is adopted in the PHY and MAC layers design. Consequently, the maximum allowed PSD for each 6-MHz channel is −47.7 dBm/Hz and −51.7 dBm/Hz for nonadjacent and adjacent channels, respectively. As a result, our HT-WBPLC system offers BPLC at least a 34-dB increase in the PSD in VHF band, which can significantly improve the achievable throughput. In mode B, the BPLC channel is

Table 7.1
HT-WBPLC Modes of Operation

	Mode A	Mode B	Mode C
Frequency Band	1.8 MHz–30 MHz	54 MHz–200 MHz	30 MHz–200 MHz
Tx Type	BPLC	BPLC + TVWS	BPLC
Rx Type	BPLC	BPLC + TVWS	BPLC + TVWS
Communication standard	HomeplugAV2	ECMA-392	HomeplugAV2
PSD	−55 dBm/Hz	−47.7 dBm/Hz (free channel), −51.7 dBm/Hz (adjacent channel)	−80 dBm/Hz
Tx/Rx Ports maximum number	2 Tx & 2 Rx	3 Tx & 3 Rx	2 Tx & 3 Rx
Number of subcarriers	1,156	1,152	1,260
Subcarrier frequency spacing	24.414 kHz	46 kHz	46 kHz

used cooperatively with the wireless TVWS channel, which increases the number of spatial channels.

7.2.1.3 Mode C

This mode deals with the case of PU presence in the VHF band. The spanned frequency band is 30 MHz–200 MHz. The HomeplugAV2 standard is adopted for the PHY and MAC layers of communication. Therefore, the PSD is restricted to −80 dBm/Hz.

The main features of the three modes of HT-WBPLC operation are summarized in Table 7.1.

7.2.2 HT-WBPLC System Model

In Figure 7.2, the system model for HT-WBPLC point-to-multipoint communication system is shown. The model is applied to the three modes of HT-WBPLC operation. The M-ary quadrature amplitude modulation (M-QAM) with gray bit mapping is used as a submodulation for the subcarriers in the OFDM symbol. Each QAM modulated signal is then sent to a space-time block coding (STBC) encoder, which transmits the same QAM symbol over both the BPLC and TVWS channels. The transmission power for each OFDM subcarrier is regulated according to the proposed standard amendment in this section, while the subcarrier and power allocation problem for MIMO HT-WBPLC channel is discussed in detail in Section 7.4.

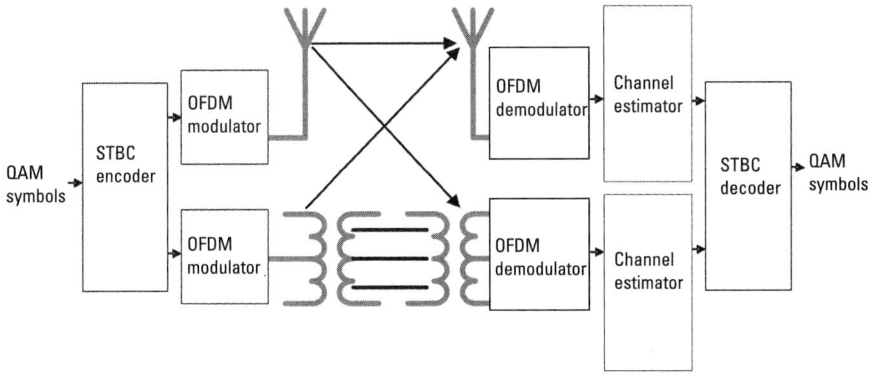

Figure 7.2 Block diagram of the HT-WBPLC system.

7.3 HT-WBPLC MIMO Channel Model

7.3.1 Channel Estimation

HT-WBPLC channel estimation is subdivided into two categories: PU and SU channel estimation. PU channel estimation is obtained using two methods: geolocation database communication and carrier sensing. Each method has its own advantages and drawbacks. Geolocation database communication supports a safer mechanism for the TV band licensed users to protect their network [10,11]. As a result, the Office of Communications (Ofcom) in United Kingdom stated "the most important mechanism in the short to medium term will be geolocation" [11]. However, the geolocation database can suffer some prediction errors and differences from real measured PU data as stated in [12], which raises the problems of PU carrier misdetection and false alarm. Hence, sensing techniques are recommended to enhance PU carrier detection. The main challenge of PU carrier sensing is determining the accurate threshold with respect to misdetection and false alarm probabilities [10]. Determining the optimum threshold for PU carrier sensing was approached in our work [13], where the throughput of point-to-point HT-WBPLC system was investigated compared to MIMO BPLC. In [13], a MIMO sensing algorithm was proposed for the PU signal, where probabilities of detection and false alarm of 0.99 and 1e-7 were achieved, respectively.

Regarding SU channel estimation, feedback is assumed between HT-WBPLC Tx and Rx. The impact of channel estimation error on the overall throughput of point-to-point HT-WBPLC system was investigated in our work in [14].

In this chapter, we focus on coverage area and throughput analysis of point-to-multipoint HT-WBPLC system compared to MIMO BPLC system in [2], ignoring the channel estimation problems for both PU and SU that had been addressed in previous literature [13,14]. Perfect channel state information is assumed available for SU Tx and Rx as in [7–9] and [15]. We rely on geolocation database communication for PU presence as recommended in [10] and [11].

7.3.2 Channel Model

BPLC indoor channel was investigated in the VHF band in [16]. The path loss of BPLC was proved to be frequency selective and dependent on several factors like power line cable type, power outlet density, power line cable branch lengths, and terminal loads. VHF path loss model for indoor wireless channel was modeled in [17]. The wireless path loss depends mainly on separating geometric distance and the number of separating walls and floors. Moreover, several studies in [18–20] suggested enhancing BPLC signal reception by adding an antenna to the receiver, which can compromise the radiated electric field.

In this chapter, we extend the channel measurements conducted in [21] and [16], to include the cross talk between BPLC and TVWS wireless channels. The aim of channel measurements is path loss modeling of the cross-talk channel and comparing it against BPLC channel path loss. In [21], BPLC wireless channel measurements in the frequency band 1.7 MHz–100 MHz show channel capacities of 450 Mbps and 85 Mbps for both short and long distances, respectively. This result proves the promising application of the standard amendment proposed in this chapter to regulate HT-WBPLC system. Since HT-WBPLC is proposed to span the VHF band up to 200 MHz, the frequency band of our channel measurements is 84 MHz–200 MHz. However, we avoid the frequency modulation (FM) band from 88 MHz–110 MHz due to interference from FM radio. The channel measurements are held inside the laboratories and offices within the Department of Electrical Engineering and Electronics at the University of Liverpool, which makes the results applicable to an indoor office environment. The Tx and Rx are located on the same floor. The channel between any two power line couplers is represented by a network of power line cables, while the channel between any two wireless antennas is the radio propagation channel. The path loss is measured using radio frequency (RF) signal generator at Tx and spectrum analyzer at Rx. The path loss is measured for three different channels: (1) the channel between two power line coupling circuits, referred to as h_{11}, (2) the channel between a coupling circuit at Tx and a wireless antenna at Rx, referred to as h_{12}, and (3) the channel between a wireless antenna at Tx and a power line coupling circuit at Rx, referred to as h_{21}.

Two broadband coupling circuits have been used in Tx and Rx. The coupling circuit has been designed to be broadband inductive as in [22] to achieve flat gain in broadband application. The loss has been measured in the coupling circuits and the connection cables by measuring their S-parameters using the network analyzer for calibration purpose. The channel measurements have been carried out between the outlets belonging to the same phase and the same distribution

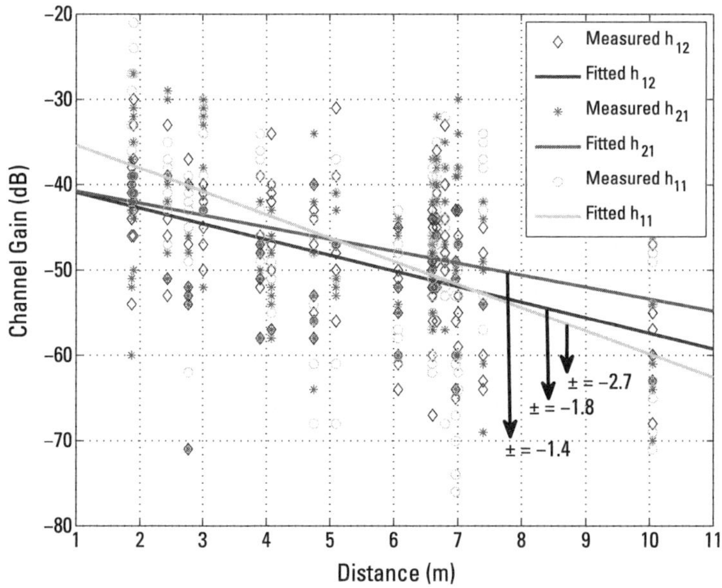

Figure 7.3 Cross talk channel compared to BPLC channel.

box. The channel measurements have been done in different rooms to have an average path loss representation.

In Figure 7.3, the gain of the cross-talk and BPLC channels are shown versus distance. The channel gain is measured at four frequencies in the band 84 MHz–200 MHz, which are 84 MHz, 110 MHz, 140 MHz, and 190 MHz. The gain in Figure 7.3 is mapped for all aforementioned four frequencies to better describe the path loss in the whole frequency band rather than single frequency tone. The gain is measured for each frequency at different coverage distances to model the path loss of the channel. The channel measurement results show that BPLC and cross-talk channels have close gain performance with respect to distance, since the slope α of the fitted channel gain takes the values of -2.7, -1.8, and -1.4 for the h_{11}, h_{12}, and h_{21}, respectively. Also, the results show that the cross-talk channel gain is below BPLC channel gain for small coverage distance; however, for long coverage distances the cross-talk channel gain outperforms the corresponding BPLC channel gain. This is due to the dependence of the cross-talk channel gain mainly on the geometric separation distance between Tx and Rx. However, BPLC channel gain is mainly dependent on the electric separation distance, which is the length of the power line cable connecting the Tx to the Rx. It is known that the electric distance is longer than the geometric distance as reported in [16].

An important observation can be deduced from Figure 7.3, which is measured channel gain large variance about its mean value. BPLC channel gain with respect to geometric Tx-Rx distance is characterized by its large variance, as have been proved in previous literature measurements [23]. This is due to BPLC channel gain dependency on Tx-Rx electric distance rather than geometric distance. As well, BPLC channel gain varies according to powerline topology, cable connections, connected loads, and outlet density, which increases the overall channel variance about its mean value.

7.4 Throughput Maximization and Power Allocation in HT-WBPLC

In this section, the downlink is investigated for a point-to-multipoint HT-WBPLC indoor system. The sink is connected to a geolocation database to obtain the PU temporal and geographical map of access to the TV channels. The noise is assumed to have zero mean and PSD N_0.

7.4.1 Problem Formulation

In HT-WBPLC, there are N_1 subcarriers in HF band 1.8 MHz–30 MHz. In this band, the HT-WBPLC transmitter uses the N_1 subcarriers of BPLC channel only under the PSD constraint of HomeplugAV2 [2]. In the VHF band 30 MHz–200 MHz, there are N_2 subcarriers that are used by HT-WBPLC transmitter under one of the following two conditions:

1. *PU absence*: In this case, HT-WBPLC transmitter uses both TVWS wireless channel and BPLC channel in the transmission of the data across the N_2 subcarriers. STBC is adopted in the transmission across the two channels in order to enhance the diversity gain. The transmission power is also constrained according to ECMA-392 standard in [3].
2. *PU presence*: In this case, HT-WBPLC transmitter uses BPLC channel only in the transmission under the PSD constraint of HomeplugAV2. STBC is here adopted in the transmission across MIMO BPLC channels only.

Since we consider the case of downlink between the sink and the users, the target of the power and subcarrier allocation is to achieve maximum throughput and also satisfy different power constraints by different standards at different frequency bands. Let $P_{e_{nj}}$, $2^{k_{nj}}$, P_{nj} be the BER, the constellation size and the allocated power for the nth subcarrier of the jth user, respectively, and N_t, N_r, ζ be the numbers of Tx and Rx ports and STBC code rate, respectively. Let $h_{n_t n_r j}$ be the channel gain of spatial path $n_t n_r$ for a given subcarrier and user and $\alpha_{n_t j}$ be the ratio of the power allocated at Tx for each spatial path. Following [24], $\alpha_{n_t j}$ is calculated as

$$\alpha_{n_t j} = \sqrt{\frac{\beta_{n_t j}}{\sum_{n_t=1}^{N_t} \beta_{n_t j}}} \tag{7.1}$$

where $\beta_{n_t j} = \sum_{n_r=1}^{N_r} |h_{n_t n_r j}|^2$. Let C_{nj} be the effective instantaneous MIMO channel gain for the nth subcarrier and the jth user after STBC decoder. C_{nj} can be calculated as $C_{nj} = \sum_{n_t=1}^{N_t} \alpha_{n_t j} \sum_{n_r=1}^{N_r} |h_{n_t n_r j}|^2$. In the case of SISO channel, C_{nj} is simply the channel gain $|h_{n_t n_r j}|^2$ and $\alpha_{n_t j}$ is taken as 1. Hence, a simple approximated probability of error expression can be adopted as in [25]:

$$P_{e_{nj}} = \frac{0.2}{\left[1 + \frac{1.6}{2^{k_{nj}}-1} \frac{P_{nj} C_{nj}}{N_t \zeta N_0}\right]^{N_t N_r}} \tag{7.2}$$

Let the $P_{e_{nj}}$ be the same for all subcarriers and users and equal to the target probability of error P_e. Therefore, the number of bits k_{nj} assigned for subcarrier n and user j is expressed as

$$k_{nj} = \log_2 \frac{1 + 1.6 P_{nj} C_{nj}}{N_t \zeta N_0 \left[(0.2/P_e)^{\frac{1}{N_t N_r}} - 1\right]} \tag{7.3}$$

The second derivative of k_{nj} with respect to P_{nj} is negative, which proves that (7.3) is a concave function [26]. k_{nj} has three different expressions according to corresponding HT-WBPLC mode of operation.

7.4.1.1 Mode A

In Mode A, the BPLC channel is the only channel available and hence, N_{t_A}, N_{r_A}, and ζ are assumed to be equal to 1 for WBPLC. However, in HT-WBPLC, N_{t_A} and N_{r_A} are equal to 2. The $P_{n_A j}$ allocated to each subcarrier must be lower than a maximum power P_A, which is the maximum power that can be allocated to the subcarrier to satisfy PSD constraint of HomeplugAV2 in HF band. The number of bits $k_{n_A j}$ assigned to mode A subcarrier n_A of jth user can be expressed using (7.3) as

$$k_{n_A j} = \log_2 \frac{1 + 1.6 P_{n_A j} C_{n_A j}}{N_{t_A} N_0 \left[(0.2/P_e)^{\frac{1}{N_{t_A} N_{r_A}}} - 1\right]} \tag{7.4}$$

7.4.1.2 Mode B: 30 MHz–84 MHz and PU Absence

In Mode B, both BPLC and TVWS channels are used. Let PH_0 represent the probability of the PU absence. Since the frequency band from 30 MHz to 54 MHz is not allowed for TVWS communication, PH_0 is equal to zero in this band. The number of bits $k_{n_B j}^{(0)}$ assigned to mode B subcarrier n_B of the jth user can be expressed as

$$k_{n_B j}^{(0)} = PH_0 \log_2 \frac{1 + 1.6 P_{n_B j}^{(0)} C_{n_B j}^{(0)}}{N_{t_B}^{(0)} N_0 \left[(0.2/P_e)^{\frac{1}{N_{t_B}^{(0)} N_{r_B}^{(0)}}} - 1\right]} \tag{7.5}$$

The allocated power $P^{(0)}_{nBj}$ for each subcarrier must be below a maximum power $P^{(0)}_{B}$ to satisfy Tx power constraint of ECMA-392 standard. $N^{(0)}_{tB}$ and $N^{(0)}_{rB}$ take the value of 2 in the case of WBPLC and the values of 2 and 3, respectively in the case of HT-WBPLC. ζ is taken as 1 for full rate STBC coding, where the superscript (0) indicates the absence of the PU.

7.4.1.3 Mode C: 30 MHz–84 MHz and PU Presence

In Mode C, the BPLC channel is the only channel used and hence, the $k^{(1)}_{nBj}$ can be expressed as

$$k^{(1)}_{nBj} = PH_1 \log_2 \frac{1 + 1.6P^{(1)}_{nBj}C^{(1)}_{nBj}}{N^{(1)}_{tB}(N_0 + N_p)\left[(0.2/P_e)^{\frac{1}{N^{(1)}_{tB}N^{(1)}_{rB}}} - 1\right]} \tag{7.6}$$

where PH_1 and N_p represent PU presence probability and interference power, respectively. The $P^{(1)}_{nBj}$ must be below a certain power $P^{(1)}_{B}$ to satisfy PSD of HomeplugAV2 in the VHF band. Let T be the overall throughput, which can be defined as

$$T = \Pi_{I0}(1 - P_e)\sum_{j=1}^{K}\left[\sum_{n_1=1}^{N_1}\frac{k_{nAj}}{t_1} + \frac{\sum_{nB,nB=1}^{N_2}(k^{(0)}_{nBj} + k^{(1)}_{nBj})}{t_2}\right] \tag{7.7}$$

where Π_{I0} is the steady-state probability of nonburst impulsive state $I0$ of the BPLC channel [27]. T is considered as linear combination of k_{nj}, which is a concave function with respect to P_{nj}. As a result, T is considered as a concave function with respect to P_{nj}. According to [26], maximizing a concave function is considered a convex optimization problem. Hence, the problem of maximizing overall throughput T can be expressed as

$$\max_{P_{nAj}, P^{(0)}_{nBj}, P^{(1)}_{nBj}} \quad T \tag{7.8}$$

s.t.

(C1) $P_{nAj}, P^{(0)}_{nBj}, P^{(1)}_{nBj} \geq 0$, (C2) $\sum_{j=1}^{K} P_{nAj} \leq P_A$,

(C3) $\sum_{j=1}^{K} P^{(0)}_{nBj} \leq P^{(0)}_{B}$, (C4) $\sum_{j=1}^{K} P^{(1)}_{nBj} \leq P^{(1)}_{B}$,

(C5) $\sum_{j=1}^{K}\left[\sum_{nA=1}^{N_1} P_{nAj} + \sum_{nB=1}^{N_2} PH_0P^{(0)}_{nBj} + PH_1P^{(1)}_{nBj}\right] \leq P_{in}$,

(C6) $\frac{\sum_{nA=1}^{N_1} k_{nAj}}{t_1} + \frac{\sum_{nB=1}^{N_2}(k^{(0)}_{nBj}+k^{(1)}_{nBj})}{t_2} \geq R_j$,

where N_1 and N_2 are the number of subcarriers in HF and VHF bands, respectively. Also, t_1 and t_2 are OFDM symbol durations in HomeplugAV2 and ECMA-392 standards, respectively. R_j is the throughput requested for each user. P_{in} is the total input power to the sink.

7.4.2 Problem Solution

The Lagrangian of the problem can be expressed as

$$
\begin{aligned}
L = & \Pi_{I0}(1-P_e)\sum_{j=1}^{K}\Bigg[\sum_{n_A=1}^{N_1}\frac{1}{t_1}\log_2\{1+\frac{1.6P_{n_Aj}C_{n_Aj}}{N_0\big[(\frac{0.2}{P_e})^{\frac{1}{(N_{tA}N_{rA})}}-1\big]}\} \\
& +\sum_{n_B}^{N_2}\frac{PH_0}{t_2}\log_2\{1+\frac{1.6P^{(0)}_{n_Bj}C^{(0)}_{n_Bj}}{(2N_0\big[(\frac{0.2}{P_e})^{\frac{1}{(N^{(0)}_{tB}N^{(0)}_{rB})}}-1\big])}\} \\
& +\frac{PH_1}{t_2}\log_2\{1+\frac{1.6P^{(1)}_{n_Bj}C^{(1)}_{n_Bj}}{2(N_0+N_p)\big[(\frac{0.2}{P_e})^{\frac{1}{(N^{(1)}_{tB}N^{(1)}_{rB})}}-1\big]}\}\Bigg] \\
& -\lambda\sum_{j=1}^{K}\Bigg[\sum_{n_A=1}^{N_1}P_{n_Aj}+\sum_{n_B=1}^{N_2}PH_0P^{(0)}_{n_Bj}+PH_1P^{(1)}_{n_Bj}\Bigg] \\
& +\lambda P_{in}-\sum_{n_A=1}^{N_1}\mu_{n_A}\Bigg[\sum_{j=1}^{K}P_{n_Aj}-P_A\Bigg] \\
& -\sum_{n_B=1}^{N_2}\mu^{(0)}_{n_B}\Bigg[\sum_{j=1}^{K}P^{(0)}_{n_Bj}-P^{(0)}_{B}\Bigg] \\
& -\sum_{n_B=1}^{N_2}\mu^{(1)}_{n_B}\Bigg[\sum_{j=1}^{K}P^{(1)}_{n_Bj}-P^{(1)}_{B}\Bigg] \\
& +\sum_{j=1}^{K}\beta_j\Bigg[\sum_{n_A=1}^{N_1}\frac{1}{t_1}\log_2\{1+\frac{1.6P_{n_Aj}C_{n_Aj}}{N_0\big[(\frac{0.2}{P_e})^{\frac{1}{(N_{tA}N_{rA})}}-1\big]}\} \\
& +\sum_{n_B=1}^{N_2}\frac{PH_0}{t_2}\log_2\{1+\frac{1.6P^{(0)}_{n_Bj}C^{(0)}_{n_Bj}}{(2N_0\big[(\frac{0.2}{P_e})^{\frac{1}{(N^{(0)}_{tB}N^{(0)}_{rB})}}-1\big])}\} \\
& +\frac{PH_1}{t_2}\log_2\{1+\frac{1.6P^{(1)}_{n_Bj}C^{(1)}_{n_Bj}}{2(N_0+N_p)\big[(\frac{0.2}{P_e})^{\frac{1}{(N^{(1)}_{tB}N^{(1)}_{rB})}}-1\big]}\}\Bigg] \\
& -\sum_{j=1}^{K}\beta_jR_j
\end{aligned}
\tag{7.9}
$$

where λ, μ_{n_A}, $\mu_{n_B}^{(0)}$, $\mu_{n_B}^{(1)}$, and β_j are Lagrangian multipliers. Using the primal decomposition method, the problem in (7.9) can be divided into three convex optimization subproblems. Each subproblem is further decomposed into single variable convex optimization subproblems as a function of the allocated power to each user and subcarrier as

Subproblem 1

$$L_1 = \Pi_{I0}(1 - P_e)\sum_{j=1}^{K}\sum_{n_A=1}^{N_1}\frac{1+\beta_j}{t_1}\log_2\left\{1+\frac{1.6P_{n_Aj}C_{n_Aj}}{N_0\left[(0.2/P_e)^{\frac{1}{N_{t_A}N_{r_A}}}-1\right]}\right\}$$
$$-\sum_{j=1}^{K}\sum_{n_A=1}^{N_1}\lambda P_{n_Aj}-\sum_{n_A=1}^{N_1}\sum_{j=1}^{K}\mu_{n_A}\left[P_{n_Aj}-\frac{P_A}{K}\right] \quad (7.10)$$

The single variable convex subproblem is represented as

$$f_1(P_{n_Aj}) = \Pi_{I0}(1 - P_e)\left[\frac{1+\beta_j}{t_1}\log_2\left\{1+\frac{1.6P_{n_Aj}C_{n_Aj}}{N_0\left[(0.2/P_e)^{\frac{1}{N_{t_A}N_{r_A}}}-1\right]}\right\}\right.$$
$$\left.-\lambda P_{n_Aj}-\mu_{n_A}\left[P_{n_Aj}-\frac{P_A}{K}\right]\right] \quad (7.11)$$

Subproblem 2

$$L_2 = \Pi_{I0}(1 - P_e)$$
$$\sum_{j=1}^{K}\sum_{n_B=1}^{N_2}\frac{PH_0(1+\beta_j)}{t_2}\log_2\left\{1+\frac{1.6P_{n_Bj}^{(0)}C_{n_Bj}^{(0)}}{(2N_0\left[(\frac{0.2}{P_e})^{\frac{1}{N_{t_B}^{(0)}N_{r_B}^{(0)}}}-1\right])}\right\}$$
$$-\sum_{j=1}^{K}\sum_{n_B=1}^{N_2}\lambda PH_0P_{n_Bj}^{(0)}-\sum_{n_B=1}^{N_2}\sum_{j=1}^{K}\mu_{n_B}^{(0)}\left[P_{n_Bj}^{(0)}-\frac{P_B^{(0)}}{K}\right] \quad (7.12)$$

The single variable convex subproblem is represented as

$$f_2(P_{n_Bj}^{(0)}) = \Pi_{I0}(1 - P_e)$$
$$\left[\frac{PH_0(1+\beta_j)}{t_2}\log_2\left\{1+\frac{1.6P_{n_Bj}^{(0)}C_{n_Bj}^{(0)}}{(2N_0\left[(\frac{0.2}{P_e})^{\frac{1}{N_{t_B}^{(0)}N_{r_B}^{(0)}}}-1\right])}\right\}\right.$$
$$\left.-\lambda PH_0P_{n_Bj}^{(0)}-\left[P_{n_Bj}^{(0)}-\frac{P_B^{(0)}}{K}\right]\right] \quad (7.13)$$

Subproblem 3

$$L_3 = \Pi_{I0}(1 - P_e)$$
$$\sum_{j=1}^{K}\sum_{n_B=1}^{N_2} \frac{PH_1(1+\beta_j)}{t_2}\log_2\left\{1 + \frac{1.6P^{(1)}_{nBj}C^{(1)}_{nBj}}{(2(N_0+N_p)[(\frac{0.2}{P_e})^{\frac{1}{(N^{(1)}_{tB}N^{(1)}_{rB})}} - 1])}\right\}$$
$$-\sum_{j=1}^{K}\sum_{n_B=1}^{N_2} \lambda PH_1 P^{(1)}_{nBj} - \sum_{n_B=1}^{N_2}\sum_{j=1}^{K} \mu^{(1)}_{nB}[P^{(1)}_{nBj} - \frac{P^{(1)}_B}{K}] \quad (7.14)$$

The single variable convex subproblem is represented as

$$f_3(P^{(1)}_{nBj}) = \Pi_{I0}(1 - P_e)$$
$$\left[\frac{PH_1(1+\beta_j)}{t_2}\log_2\left\{1 + \frac{1.6P^{(1)}_{nBj}C^{(1)}_{nBj}}{(2(N_0+N_p)[(\frac{0.2}{P_e})^{\frac{1}{(N^{(1)}_{tB}N^{(1)}_{rB})}} - 1])}\right\}\right.$$
$$\left. - \lambda PH_1 P^{(1)}_{nBj} - \mu^{(1)}_{nB}[P^{(1)}_{nBj} - \frac{P^{(1)}_B}{K}]\right] \quad (7.15)$$

after forming the Lagrangian functions and applying the Karush-Kuhn-Tucker (KKT) conditions. The power allocated to each mode of operation for subcarrier n and user j is expressed as

$$P_{nAj} = \left[\frac{1+\beta_j}{(t_1(\lambda + \mu_{nA}))} - \frac{N_0[(\frac{0.2}{P_e})^{\frac{1}{(N_{tA}N_{rA})}} - 1]}{1.6C_{nAj}}\right] \quad (7.16)$$

$$P^{(0)}_{nBj} = \left[\frac{PH_0(1+\beta_j)}{(t_2(PH_0\lambda + \mu^{(0)}_{nB}))} - \frac{2N_0[(\frac{0.2}{P_e})^{(\frac{1}{(N^{(0)}_{tB}N^{(0)}_{rB})})} - 1]}{1.6C^{(0)}_{nBj}}\right] \quad (7.17)$$

$$P^{(1)}_{nBj} = \left[\frac{PH_1(1+\beta_j)}{(t_2(PH_1\lambda + \mu^{(1)}_{nB}))} - \frac{2(N_0+N_p)[(\frac{0.2}{P_e})^{(\frac{1}{N^{(1)}_{tB}N^{(1)}_{rB}})} - 1]}{1.6C^{(1)}_{nBj}}\right] \quad (7.18)$$

Using (7.16)–(7.18), the power and subcarriers are allocated for each user according to Algorithm 1.

7.5 Numerical Results

In this section, the HT-WBPLC downlink model in Figure 7.2 is used for simulation. The PU presence activity is represented using two-state discrete-time

Algorithm 1 Subcarrier and user allocation

Initialization: Set $\phi = \{1 \dots K\}$ as the set of users, $\Omega_T = \{1 \dots N_1\} \cup \{1 \dots N_2\}$ as the set of all available subcarriers for all users. Ω_j is the set of the subcarriers allocated to user $j \in \phi$, $\Omega_j = \{\phi\}$. $\beta_j, \mu_{n_A}, \mu_{n_B}^{(0)}, \mu_{n_B}^{(1)} = 0$.
Allocate the maximum permissible power for each subcarrier $P_{n_Aj} = P_A$, $P_{nBj}^{(0)} = P_B^{(0)}, P_{nBj}^{(1)} = P_B^{(1)}$.
Start a For loop across the subcarriers of Ω_T
for k=1:N_1 **do**
    Calculate k_{n_Aj}/t_1 or $(k_{nBj}^{(0)} + k_{nBj}^{(1)})/t_2$ for each user j in ϕ.
    if $k_{n_Aj}/t_1 > k_{n_A(j-1)}/t_1$ **then**
        $j \leftarrow n_A$, $\Omega_T = \Omega_T - n_A$
    else
        if $(k_{nBj}^{(0)} + k_{nBj}^{(1)})/t_2 > (k_{nB(j-1)}^{(0)} + k_{nB(j-1)}^{(1)})/t_2$ **then**
            $j \leftarrow n_B$, $\Omega_T = \Omega_T - n_B$
        end if
    end if
    if R_j is achieved **then**
        $\phi = \phi - j$
    end if
end for
while sum of all transmission power is $\leq P_{in}$ **do**
    Calculate the optimum power using the obtained user-frequency map in the For loop with the aid of (7.16)–(B.1), and find the appropriate λ using the bisection method.
    if $P_{n_Aj} > \mu_{n_A}$ **then**
        $P_{n_Aj} \leftarrow \mu_{n_A}$
    else
        if $P_{nBj}^{(0,1)} > \mu_{nB}$ **then**
            $P_{nBj}^{(0,1)} \leftarrow \mu_{nB}$
        end if
    end if
end while

Markov chain model (DTMC), where the PU presence steady-state probability for each 6-MHz TV channel is taken as 0.2. Impulsive noise is represented in our simulations by a two-state DTMC model, where the steady-state probability of nonimpulsive state Π_{I0} is 0.9. The power line topology generator in [28] is used in the simulations to model the power line cable, terminal loads, and outlet

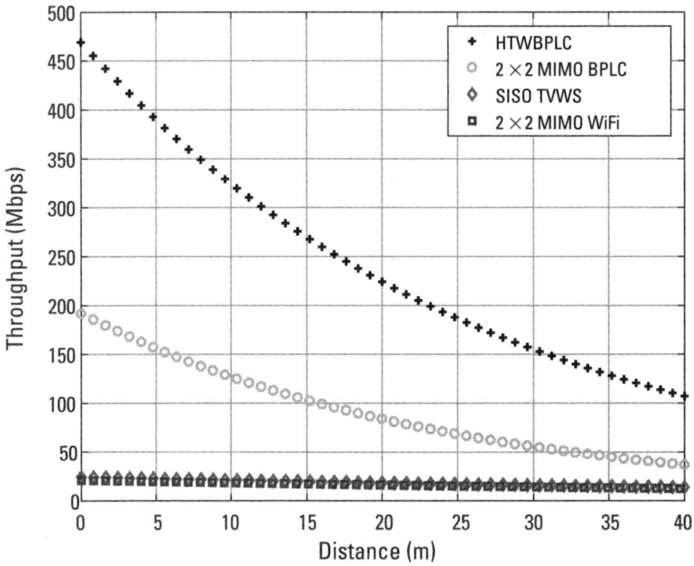

Figure 7.4 Achievable throughput versus coverage distance.

connection and hence, the power line channel. The topology generator in [28] is used to generate 100 topologies for each case of simulation and to calculate the average throughput. The VHF radio propagation channel is modelled as Rayleigh fading channel with average path loss as in [17]. In the simulations, HT-WBPLC is compared to MIMO BPLC [2] and TVWS systems. TVWS is implemented using ECMA-392 standard [3], and the FCC regulations [29].

7.5.1 HT-WBPLC Coverage

In Figure 7.4, HT-WBPLC coverage distance is compared to those of 2×2 MIMO WiFi, TVWS, and MIMO BPLC communication systems. The achievable throughput by the proposed HT-WBPLC nearly doubles that achieved by MIMO BPLC. This proves the ability of our proposed system of improving both the coverage distance and throughput over other communication technologies.

Since the coverage area is considered as a main challenge for current indoor network technologies, a comparison between our proposed HT-WBPLC system and MIMO BPLC system is presented. In the comparison, a 500-m^2 area with 20 users is simulated for different BPLC topologies, load connections, and users' distribution across the area. The average throughput for each user is calculated and compared against its location with respect to the sink location and hence, a coverage heat map is generated.

In Table 7.2 the coverage areas for HT-WBPLC and MIMO BPLC are presented, respectively. HT-WBPLC shows more ability to cover larger areas

Table 7.2
HT-WBPLC and MIMO BPLC Coverage area

Coverage area	HT-WBPLC Throughput	MIMO BPLC Throughput
25 m^2	95 mbps	70 mbps
100 m^2	95 mbps	70 mbps
225 m^2	75 mbps	55 mbps
400 m^2	60 mbps	30 mbps

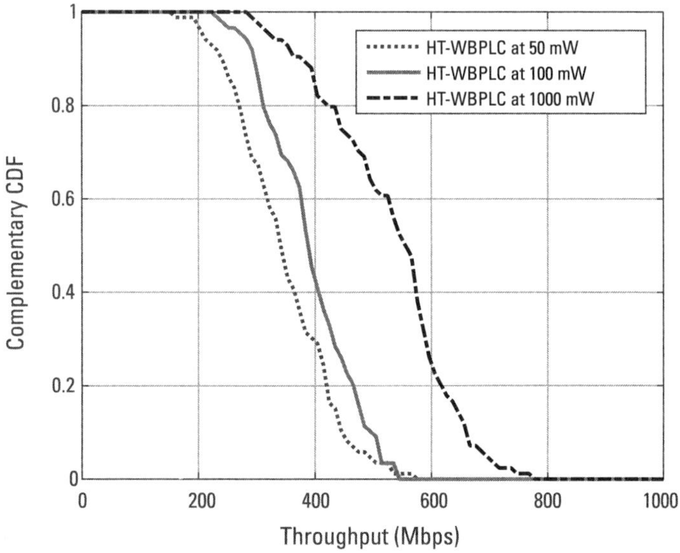

Figure 7.5 Complementary CDF of the achievable throughput by the proposed HT-WBPLC using different input power levels.

with preserving high speed compared to MIMO BPLC. The ability of the HT-WBPLC system to increase the user throughput for large coverage areas increased by nearly 29% compared to MIMO BPLC.

7.5.2 HT-WBPLC Throughput

In Figure 7.5, the throughput CCDF of HT-WBPLC system is shown at different input power levels. The input power levels simulated are 50 mW, 100 mW, and 1 W, which are allocated in the downlink for different users and different subcarriers using Algorithm 1 in Section 7.4. For 50% of the time, HT-WBPLC achieves total downlink throughput of 565 Mbps and 343 Mbps at input transmission power of 50 mW and 1000 mW, respectively. Therefore, an increase of 200 Mbps is observed in the achievable throughput with the increase in the power level. Compared to MIMO BPLC [2], the proposed HT-WBPLC can

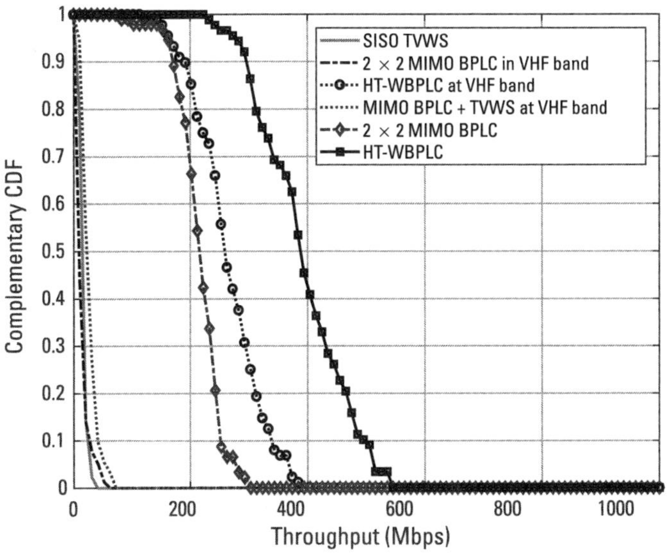

Figure 7.6 Complementary CDF of the throughput using the proposed HT-WBPLC and the conventional MIMO BPLC.

support higher input power level as it complies with the requirements of the FCC regulations, which allow a 34-dB increase in the PSD than that allowed by HomeplugAV2 for BPLC.

In Figure 7.6, our proposed HT-WBPLC is compared to MIMO BPLC in the frequency band 1.8 MHz–200 MHz at an input transmission power of 100 mW. The simulation results show that for 60% of the time, a throughput of 200 Mbps is achieved using MIMO BPLC [2], while 380 Mbps is achieved using HT-WBPLC, which corresponds to a 90% increase. The HT-WBPLC throughput CCDF is compared against MIMO BPLC [2] and TVWS in VHF band only. Significant throughput increase has been shown in VHF band by HT-WBPLC over the other two systems. This proves that the main reason behind the increase in the achievable throughput is due to VHF band exploitation. The increase is for two main reasons: (1) the addition of the wireless channel to the BPLC channel by adding a wireless VHF antenna allows HT-WBPLC to be TVWS standard compliant, which results in higher permissible PSD and hence, higher throughput compared to HomeplugAV2, and (2) the increase in allowed spectrum of communication in VHF band up to 200 MHz.

7.6 Summary

In this chapter, a HT-WBPLC communication system was been proposed, which makes use of TVWS and BPLC channels in VHF bands cooperatively based on the cognitive radio principle. HT-WBPLC benefits are twofold: (a) HT-WBPLC overcomes IoT network design challenge in realizing an integrated solution that incorporates BPLC and TVWS communication based nodes by offering a TVWS compliant perspective of BPLC standard, and (b) HT-WBPLC also overcomes PSD constraints in BPLC and enables transmission in an additional frequency band, and hence more coverage area. Our channel measurement results in Section 7.3 prove that the cross-talk channel between BPLC and TVWS can be exploited to enhance BPLC signal reception. Point-to-multipoint system simulation results have demonstrated that the proposed HT-WBPLC features a minimum of 40% over MIMO BPLC [2] and TVWS [3,29], in term of total downlink throughput. Regarding network coverage, HT-WBPLC shows an increase in achievable user throughput by 29% compared to MIMO BPLC [2] for areas larger than or equal to 400 m^2. Therefore, HT-WBPLC is a promising solution for the growing needs of IoT.

References

[1] Ferreira, H. C., L. Lampe, J. Newbury, and T. Swart, (eds.), *Power Line Communications: Theory and Applications for Narrowband and Broadband Communications over Power Lines*, Chichester, United Kingdom: John Wiley & Sons, Inc., 2010.

[2] Berger, L. T., A. Schwager, P. Pagani, and D. Schneider, *MIMO Power Line Communications: Narrow and Broadband Standards, EMC, and Advanced Processing*, Boca Raton, FL: CRC Press, 2014.

[3] MAC and PHY for Operation in TV White Space, ECMA-392, December 2009.

[4] Zhang, Y., and C. Leung, "Resource Allocation in an OFDM-based Cognitive Radio System," *IEEE Transactions on Communications*, Vol. 57, No. 7, July 2009.

[5] Saki, H. and M. Shikh-Bahaei, "Cross-Layer Resource Allocation for Video Streaming over OFDMA Cognitive Radio Networks," *IEEE Transactions on Multimedia*, Vol. 17, No. 3, March 2015, pp. 333–345.

[6] Tsiropoulos, G. I., O. A. Dobre, M. H. Ahmed, and K. E. Baddour, "Radio Resource Allocation Techniques for Efficient Spectrum Access in Cognitive Radio Networks," *IEEE Communications Surveys & Tutorials*, Vol. 18, No. 1, January 2016, pp. 824–847.

[7] Ngo, D. T., C. Tellambura, and H. H. Nguyen, "Resource Allocation for OFDMA-based Cognitive Radio Multicast Networks with Primary User Activity Consideration," *IEEE Transactions on Vehicular Technology*, Vol. 59, No. 4, May 2010, pp. 1668–1679.

[8] Zhang, S., W. Xu, S. Li, and J. Lin, "Resource Allocation for Multiple Description Coding Multicast in OFDM-based Cognitive Radio Systems with Non-Full Buffer Traffic," in *Proc. IEEE Wireless Communications and Networking Conference (WCNC)*, April 2013, Shanghai, China, pp. 199–204.

[9] Yang, K., W. Xu, S. Li, and J. Lin, "Distributed Multicast Resource Allocation in OFDM-based Cognitive Radio Networks," in *8th International ICST Conference on Communications and Networking in China (CHINACOM)*, August 2013, Guilin, China, pp. 57–62.

[10] Makris, D., G. Gardikis, and A. Kourtis, "Quantifying TV White Space Capacity: A Geolocation-Based Approach," *IEEE Communications Magazine*, Vol. 50, No. 9, September 2012, p. 145.

[11] Gurney, D., G. Buchwald, L. Ecklund, S. L. Kuffner, and J. Grosspietsch, "Geo-Location Database Techniques for Incumbent Protection in the TV White Space," in *Proc. 3rd IEEE Symposium on New Frontiers in Dynamic Spectrum Access Networks*, October 2008, Chicago, pp. 1–9.

[12] Achtzehn, A., J. Riihijarvi, and P. Mahonen, "Improving Accuracy for TVWS Geolocation Databases: Results from Measurement-Driven Estimation Approaches," in *Proc. IEEE International Symposium on Dynamic Spectrum Access Networks (DYSPAN)*, April 2014, McLean, VA, pp. 392–403.

[13] Heggo, M., X. Zhu, Y. Huang, and S. Sun, "A Hybrid Power Line and TV White Space MIMO System for Indoor Broadband Communications," in *Proc. IEEE 84th Vehicular Technology Conference (VTC) Fall*, September 2016, Montreal, Canada.

[14] Heggo, M., X. Zhu, S. Sun, and Y. Huang, "A Cognitive TV White Space-Broadband Power Line MIMO System for Indoor Communication Networks," *Journal of the Franklin Institute*, Vol. 355, No. 11, July 2018, pp. 4755–4770.

[15] Bounouader, N., G. Aniba, Z. Guennoun, et al., "Exploiting Zero Forcing Beamforming and TV White Space Band for Multiuser MIMO Cognitive Cooperative Radio Networks," in *Proc. International Conference on Wireless Networks and Mobile Communications (WINCOM)*, October 2015, Marrakech, Morocco, pp. 1–6.

[16] Heggo, M., X. Zhu, Y. Huang, and S. Sun, "A Novel Statistical Approach of Path Loss Mapping for Indoor Broadband Power Line Communications," in *Proc. IEEE International Conference on Smart Grid Communications (SmartGridComm)*, November 2014, Venice, Italy, pp. 499–504.

[17] Andrusenko, J., R. L. Miller, J. A. Abrahamson, N. M. Merheb Emanuelli, R. S. Pattay, and R. M. Shuford, "VHF General Urban Path Loss Model for Short Range Ground-to-Ground Communications," *IEEE Transactions on Antennas and Propagation*, Vol. 56, No. 10, October 2008, pp. 3302–3310.

[18] Finamore, W. A., M. V. Ribeiro, and L.Lampe, "Advancing Power Line Communication: Cognitive, Cooperative, and MIMO Communication," in *Proc. Brazilian Telecommunications Symposium*, September 2012, Brasilia, Brazil, pp. 13–16.

[19] Degauque, P., P. Laly, V. Degardin, M. Lienard, and L. Diquelou, "Compromising Electromagnetic Field Radiated by in-House PLC Lines," in *Proc. IEEE Global Telecommunications Conference (GLOBECOM)*, December 2010, Miami, Florida, pp. 1–5.

[20] Degardin, V., P. Laly, M. Lienard, and P. Degauque, "Compromising Radiated Emission from a Power Line Communication Cable," *Journal of Communications Software & Systems*, Vol. 7, No. 1, March 2011, pp. 16–21.

[21] Oliveira, T., F. Andrade, A. Picorone, H. Latchman, S. Netto, and M. Ribeiro, "Characterization of Hybrid Communication Channel in Indoor Scenario," *Journal of Communication and Information Systems*, Vol. 31, No. 1, September 2016.

[22] Chen, S., "Ultra Wideband Gigabit Powerline Communication," Ph.D. Dissertation, Queen Mary University, London, UK, 2009.

[23] Tonello, A. M., and F. Versolatto, "Bottom-up Statistical PLC Channel Modeling, Part II: Inferring the Statistics," *IEEE Transactions on Power Delivery*, Vol. 25, No. 4, October 2010, pp. 2356–2363.

[24] Xian, L., and H. Liu, "An Adaptive Power Allocation Scheme for Space-Time Block Coded MIMO Systems," in *Proc. IEEE Wireless Communications and Networking Conference*, March 2005, New Orleans, Louisiana, pp. 504–508.

[25] Wang, D., H. Minn, and N. Al-Dhahir, "A Robust Asynchronous Multiuser STBC-OFDM Transmission Scheme for Frequency- Selective Channels," *IEEE Transactions on Wireless Communications*, Vol. 7, No. 10, 2008, pp. 3725–3731.

[26] Boyd, S., and L. Vandenberghe, *Convex Optimization*, Cambridge, United Kingdom: Cambridge University Press, 2004.

[27] Yin, J., X. Zhu, and Y. Huang, "Modeling of Amplitude-Correlated and Occurrence-Dependent Impulsive Noise for Power Line Communication," in *Proc. IEEE International Conference on Communications (ICC)*, June 2014, Sydney, Australia, pp. 4565–4570.

[28] Tonello, A. M., and F. Versolatto, "Bottom-Up Statistical PLC Channel Modeling, Part I: Random Topology Model and Efficient Transfer Function Computation," *IEEE Transactions on Power Delivery*, Vol. 26, No. 2, April 2011, pp. 891–898.

[29] Federal Communications Commission, Second Report and Order and Memorandum Opinion and Order in the Matter of Unlicensed Operation in the TV Broadcast Bands, Additional Spectrum for Unlicensed Devices Below 900 MHz and in the 3 GHz Band, Document 08-260, November 2008.

Part III:

Hybrid UHF Wireless-Power Line Sensor Networks

8

Overview and Applications of Sensor Networks

The IoT originates from cross-discipline research in various fields such as electronics, telecommunication, computer networks, and embedded systems [1]. As already outlined, the basic idea is that many more devices will be able to interact and communicate with each other [2, 3]. This chapter first reviews the basic architecture of sensor networks, then discusses some applications of sensor networks.

8.1 Architecture of Sensor Networks

Sensor nodes are deployed to collect ambient conditions (e.g., temperature, humidity, sound, movement) from the observed phenomenon and then the measurements can be processed and analyzed to reveal some characteristics of interest about the surrounding environment of the sensor nodes [12]. A larger number of the sensor nodes can be networked to perform specific tasks collaboratively, thus forming a sensor network.

Typically, a sensor node consists of four basic components [13]; namely, a sensing unit, a processing unit, a communication unit, and a power unit. As shown in Figure 8.1, there are two subunits within the sensing unit, which are called the sensor and the analog-to-digital converter (ADC). The sensor collects target information from the observed environment and feeds it into the ADC in analog signals. Then the analog signals are converted by the ADC into digital signals and the digital signals are transferred to the processing unit. The processing unit, which generally contains a small storage unit, performs some local information processing, such as data encryption and compression, and manages the procedures

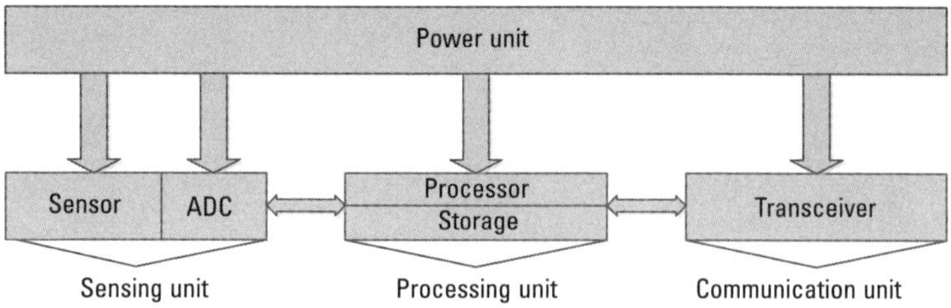

Figure 8.1 Basic components of a sensor node.

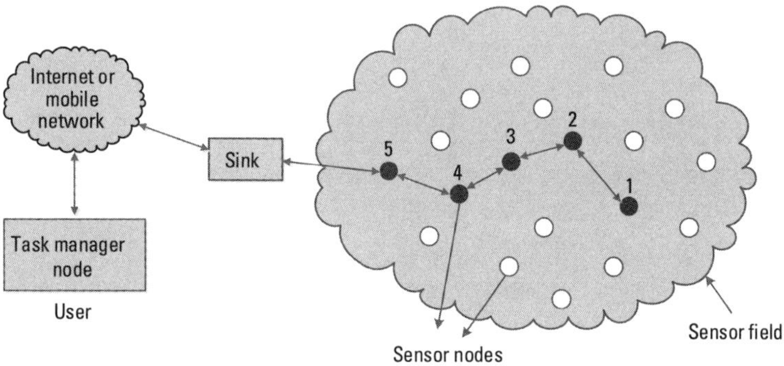

Figure 8.2 Overview of a sensor network [13].

that allow the sensor node collaborate with the other nodes to carry out assigned sensing tasks [13]. The communication unit connects the node to the sensor network. Finally, the power unit provides energy supply to all the other units. In addition, depending on the specific application, the sensor node may also contain additional units, such as a location-finding system to provide knowledge of location and a mobilizer that allows the sensor node to change its location or configuration (e.g., to change antenna's radiation direction to enhance the signal strength) [14].

A sensor network generally contains hundreds or thousands of sensor nodes [12], which are either randomly deployed (e.g., dropped from an airplane in a disaster area) or the positions of the sensor nodes are predetermined (e.g., fire alarm sensors in a facility). As shown in Figure 8.2, the sensor nodes are scattered in an area called the sensor field. Each of the sensor nodes is capable of collecting intended data and route data back to the sink node. The sink node may communicate with the task manager node via internet or mobile network such that the user can have access to the collected information and further data processing and analyzing can be carried out.

In particular, data collection and communication in sensor nodes can be divided as [15]: *clock-driven*, *event-driven*, and *query-driven*. In a *clock-driven* data collection and communication fashion, each sensor node gathers and sends data at constant periodic intervals. Over time these periodically collected data can be combined to produce temporal and spatial information about the sensor field [15]. An example of such a data gathering and communication scheme is the monitoring of the humidity of soil, where the sensor nodes are buried into the soil and "snapshots" of the humidity is collected and routed to the sink node periodically. The *event-driven* and *query-driven* data collection and communication are triggered by certain events or queries, such as the detection of fire in a warehouse or the user requires the sensor nodes to report their positions. In addition, the communications between sensor nodes to the sink node can be classified into three categories as summarized below [15].

1. *Direct transmission*: In this transmission scheme, each sensor node forwards its locally generated data directly to the sink node and there is no communication between the sensor nodes. The advantage of such a transmission scheme is simple in design. However, for a sensor network that spans a large area, this approach has an inherent scalability problem since it is a many-to-one communication [15], especially when hundreds or thousands of sensor nodes are deployed in the sensor field. Also, the distance of the sink node away from the sensor field is limited since all the sensor nodes should be in reach with the sink directly. In addition, for the sensor nodes that are far away from the sink node, they may consume more energy for transmission than those sensor nodes that are close to the sink. This means that the sensor nodes that are far away from the sink node may be drained of energy quickly. Depending on the specific applications, the data rate of each sensor node varies between several kbps to several hundred kbps. In particular, the sensor nodes may not continuously send data. Rather they may send that at intermittent intervals or on triggering of certain events. This helps in persevering the battery.
2. *Multihop routing*: In such a communication scheme, each sensor node in the sensor field plays a dual role as data transmitter and data router and may communicate with the sink node in a multihop fashion. For example, in Figure 8.2, sensor node 1 collects data and forwards it to the sink node through the path 1→2→3→4→5→*Sink*. This approach requires collaboration between sensor nodes and it can be designed to realize different objectives (e.g., maximize network throughput or minimize per node energy consumption). One potential drawback of such a routing scheme, however, is that there may exist some hot spots to which many sensor nodes will forward their data. The sensor nodes on hot spots may be exhausted of energy rapidly and cause significant topological changes and rerouting of packets and reorganization of the network may be necessary [12]. Generally in this case, depending on the specific applications, the data rate of the sensor nodes that are far

from the sink node varies between several kbps to several hundred kbps. While the sensor nodes that are closer to the sink node perform more duties on relaying and therefore have data rate on the order of several hundred kbps to several Mbps.

3. *Clustering*: In a clustered sensor network, the sensor nodes form clusters dynamically with neighboring sensors. This approach localizes traffic and allows more scalability of the sensor network by selecting a sensor node in each cluster as the cluster head that is responsible for relaying data from each sensor node in the cluster to the sink node. In addition, data fusion and compression may occur in the cluster head due to the fact that data from neighboring sensors that are close enough may be in high correlation [15]. As well, due to the nature of the cluster head, it will inevitably consume more energy and run out of power supply than the rest of the nodes in the cluster. Therefore, some protocol designs (e.g., the Low Energy Adaptive Clustering Hierarchy [16]) have focused on distributed cluster formation and dynamic cluster head selection problem. Depending on the specific applications, the data rate of each sensor node (except the cluster head) varies between several kbps to several hundred kbps, while the sensor nodes performing as the cluster heads have data rate on the order of several hundred kbps to several Mbps.

8.2 Applications of Sensor Networks

The networked sensors are widely used and deployed to collect measurements from entities of interest; for example, they can be distributed on the ground or in the soil, or embedded inside building structures or in human bodies to monitor environmental parameters and detect the occurrence of events [17]. Some applications of sensor networks are listed as follows.

1. *Structural health monitoring*: Sensor networks can be deployed to monitor structural parameters such as strain in a large region. Broadly speaking, structural health monitoring [18–21] aims at using sensor networks to localize damage that is significant enough to influence the properties of the entire structure or large sections of it (e.g., severe damage to an entire cable on a bridge) through structural responses due to external excitations such as heavy winds and passing vehicles. An example of such a sensor network is the deployment of 64 wireless sensor nodes over the main span and a tower of the Golden Gate Bridge to measure ambient structural vibrations [22].
2. *Industrial process control*: With the rapid improvement and miniaturization in hardware, low-cost hardware components such as complementary metal-oxide semiconductor (CMOS) cameras and

microphones are integrated in wireless sensor nodes, thus enabling the development of wireless multimedia sensor networks [23, 24]. Such networks are able to retrieve visual and audio content, still images, and scalar sensor data such as temperature, pressure, and location of objects. Therefore, wireless multimedia sensor networks are able to simplify and add flexibility to machine vision systems for visual inspections and automated actions in automated manufacturing processes [23].

3. *Homeland security*: Electrochemical sensor networks can be used to detect weapons of mass destruction such as the bioweapons equipped by terrorists [17]. Sensor networks can be used to monitor and protect critical infrastructures such as railways, airports, utility, and water supplies [25]. For example, sensor networks can be deployed in the airport to monitor the chemistry of the environment such as identifying toxic odors.
4. *Healthcare*: Wireless sensor nodes can be placed inside, on and around the human body to form a body area network (BAN) so that physiological signs and parameters such as blood pressure can be monitored [26]. Unlike conventional wired healthcare system, the wireless body area network (WBAN) potentially provides ubiquitous real-time monitoring without compromising the convenience of users [17].

References

[1] Atzori, L., A. Iera, and G. Morabito, "The Internet of Things: A Survey," *Computer Networks*, Vol. 54, No. 15, 2010, pp. 2787–2805.

[2] Zanella, A., N. Bui, A. Castellani, L. Vangelista, and M. Zorzi, "Internet of Things for Smart Cities," *IEEE Internet of Things Journal*, Vol. 1, No. 1, 2014, pp. 22–32.

[3] Yaqoob, I., E. Ahmed, I. A. T. Hashem, et al., "Internet of Things Architecture: Recent Advances, Taxonomy, Requirements, and Open Challenges," *IEEE Wireless Communications*, Vol. 24, No. 3, 2017, pp. 10–16.

[4] Fadel, E., V. C. Gungor, L. Nassef, et al., "A Survey on Wireless Sensor Networks for Smart Grid," *Computer Communications*, Vol. 71, 2015, pp. 22–33.

[5] Bellavista, P., G. Cardone, A. Corradi, and L. Foschini, "Convergence of MANET and WSN in IoT Urban Scenarios," *IEEE Sensors Journal*, Vol. 13, No. 10, 2013, pp. 3558–3567.

[6] Hu, L., M. Qiu, J. Song, M. Shamim Hossain, and A. Ghoneim, "Software Defined Healthcare Networks," *IEEE Wireless Communications*, Vol. 22, No. 6, 2015, pp. 67–75.

[7] Samanta, A., S. Bera, and S. Misra, "Link-Quality-Aware Resource Allocation with Load Balance in Wireless Body Area Networks," *IEEE Systems Journal*, Vol. 12, No. 1, 2015, pp. 74–81.

[8] Pan, J., R. Jain, S. Paul, T. Vu, A. Saifullah, and M. Sha, "An Internet of Things Framework for Smart Energy in Buildings: Designs, Prototype, and Experiments," *IEEE Internet of Things Journal*, Vol. 2, No. 6, 2015, pp. 527–537.

[9] Bera, S., S. Misra, and J.J. P. C. Rodrigues, "Cloud Computing Applications for Smart Grid: A Survey," *IEEE Transactions on Parallel and Distributed Systems*, Vol. 26, No. 5, 2015 1477–1494.

[10] Iyengar, S. S., and R. R. Brooks, "Distributed Sensor Networks: Sensor Networking and Applications," Boca Raton, FL: CRC Press, 2016.

[11] Al-Fuqaha, A., M. Guizani, M. Mohammadi, M. Aledhari, and M. Ayyash, "Internet of Things: A Survey on Enabling Technologies, Protocols, and Applications," *IEEE Communications Surveys & Tutorials*, Vol. 17, No. 4, 2015, pp. 2347–2376.

[12] Al-Karaki, J. N., and A. E. Kamal, "Routing Techniques in Wireless Sensor Networks: A Survey," *IEEE Wireless Communications*, Vol. 11, No. 6, 2004, pp. 6–28.

[13] Akyildiz, I. F., W. Su, Y. Sankarasubramaniam, and E. Cayirci, "A Survey on Sensor Networks," *IEEE Communications Magazine*, Vol. 40, No. 8, 2002, pp. 102–114.

[14] Anastasi, G., M. Conti, M. Di Francesco, and A. Passarella, "Energy Conservation in Wireless Sensor Networks: A Survey," *Ad Hoc Networks*, Vol. 7, No. 3, 2009, pp. 537–568.

[15] Duarte-Melo, E. J., and M. Liu, "Analysis of Energy Consumption and Lifetime of Heterogeneous Wireless Sensor Networks," in *Proc. IEEE Global Communications Conference (GLOBECOM)*, Taipei, Taiwan, November 2002, Vol. 1, pp.21–25.

[16] Rabiner Heinzelman, W., A. Chandrakasan, and H. Balakrishnan, "Energy-Efficient Communication Protocol for Wireless Microsensor Networks," in *Proc. IEEE Annual Hawaii International Conference on System Sciences*, Hawaii, January 2000.

[17] Shuguang, R. C., "Cross-Layer Optimization in Energy Constrained Networks," Ph.D. Thesis, Stanford University, 2005.

[18] Chintalapudi, K., T. Fu, J. Paek, et al., "Monitoring Civil Structures with a Wireless Sensor Network," *IEEE Internet Computing*, Vol. 10, No. 2, 2006, pp. 26–34.

[19] Harms, T., S. Sedigh, and F. Bastianini, "Structural Health Monitoring of Bridges Using Wireless Sensor Networks," *IEEE Instrumentation & Measurement Magazine*, Vol. 13, No. 6, 2010.

[20] Wang, F., D. Wang, and J. Liu, "High-Rise Structure Monitoring with Elevator-Assisted Wireless Sensor Networking: Design, Optimization, and Case Study," *Wireless Networks*, Vol. 25, 2017, pp. 1–19.

[21] Hackmann, G., W. Guo, G. Yan, Z. Sun, C. Lu, and S. Dyke, "Cyber-Physical Codesign of Distributed Structural Health Monitoring with Wireless Sensor Networks," *IEEE Transactions on Parallel and Distributed Systems*, Vol. 25, No. 1, 2014, pp. 63–72.

[22] Sukun Kim, Shamim Pakzad, David Culler, , et al., "Health Monitoring of Civil Infrastructures Using Wireless Sensor Networks," in *Proc. ACM International Conference on Information Processing in Sensor Networks*, Cambridge, Massachusetts, 2007, pp. 254–263.

[23] Akyildiz, I. F., T. Melodia, and K. R. Chowdury, "Wireless Multimedia Sensor Networks: A Survey," *IEEE Wireless Communications*, Vol. 14, No. 6, 2007.

[24] Lin, K., J. J. P. C. Rodrigues, H. Ge, N. Xiong, and X. Liang, "Energy Efficiency QOS Assurance Routing in Wireless Multimedia Sensor Networks," *IEEE Systems Journal*, Vol. 5, No. 4, pp. 495–505, 2011.

[25] Lee, K. B., and M. E. Reichardt, "Open Standards For Homeland Security Sensor Networks," *IEEE Instrumentation & Measurement Magazine*, Vol. 8, No. 5, 2005, pp. 14–21.

[26] Patel, M., and J. Wang, "Applications, Challenges, and Prospective in Emerging Body Area Networking Technologies," *IEEE Wireless Communications*, Vol. 17, No. 1, 2010.

9

Cross-Layer Network Lifetime Maximization for Hybrid Sensor Networks

In this chapter, a hybrid sensor network, which consists of both wireless sensor nodes and PL sensor nodes, is proposed for industrial sensor network applications. In particular, the network lifetime is chosen as the main design criteria to demonstrate the performance improvement of such a hybrid sensor network as compared to the traditional pure WSNs. This study can also be applied to other applications of the sensor network with stringent energy budgets, such as structural health monitoring, where building stress and motion sensors are inserted into the concrete before it is poured [1].

9.1 Real-Life Application

The next-generation industrial automation system involves three different levels [2]: (1) the field level, where the automation process is monitored and controlled directly by the sensors and actuators (2) the automation level, where the industrial controllers, such as programmable logic controllers are used to perform the process control decision making, (3) and the management level, where best-effort IP traffic is exchanged. Typically, the devices in the field level are interconnected by an industrial wireless sensor network, while the automation and management levels are connected to wired networks [3].

In such networks, the wireless sensors in the automation level are required to constantly monitor and sample the process and to send critical messages such as the measurement and actuation signals to the automation and management levels within a given time. This system ensures that a process disturbance or emergency can be handled in a timely manner; otherwise, actuators may misbehave and potentially cause material or physical damage or even hazardous consequences. An example of such a network is in the chemical and petroleum refining industry [4] where wireless sensors are scatted in the chemical reaction pool or refineries to monitor the process. For a wired network in automation and management levels, the cost of installing wires for communication within refineries is very high due to safety requirements [5]. One potential candidate for the implementation of a wired network in industrial applications is the PLC, which has the advantage of the ubiquitous infrastructure of PL cables. As well, in situations where wired communication infrastructure is a problem (e.g., installing new communication systems in old facilities), PLC is an attractive solution that provides a much lower installation time and costs [6].

9.2 Chapter Overview

In this chapter, a hybrid sensor network for industrial automation systems, that consist of both wireless and PL sensor nodes is proposed. This work is different in the following aspects. First, to the best of our knowledge, it is the first reported work in the literature that focuses on the cross-layer design of such a heterogeneous network. The hybrid sensor network takes advantage of the flexibility of WSNs while the PL sensors are deployed to prolong the lifetime of the network. This work studies the joint design of the PHY, MAC, and network layers to maximize the hybrid network lifetime, which is limited by the battery capacity of wireless sensors. Second, closed-form expressions of a globally optimal solution for lifetime maximization of a hybrid sensor network are derived for the linear topology. Such closed-form solutions give insights into factors that are significant to the network lifetime when designing a hybrid sensor network. Third, the impacts of different network configurations such as source rate and sensor node densities on a hybrid network lifetime are investigated. Finally, the impact of different transmission strategies of PL nodes on the effectiveness of the network is studied.

The rest of this chapter is organized as follows. Related work is presented in Section 9.3. Section 9.4 describes the system model. Section 9.5 formulates the optimization problem. In Section 9.6, analytical expressions for the hybrid network lifetime are derived for the linear topology. Section 9.7 analyzes the numerical results. This chapter is summarized in Section 9.8.

9.3 Related Work

The ubiquitous deployment of WSNs is limited by the energy supply of wireless sensor nodes since energy is a scarce resource. This has caused a tremendous upsurge of research interest on prolonging the network lifetime of WSNs through different approaches. Network lifetime has various definitions in the research community. For example, in [7–10], the network lifetime is defined as the the time duration until the first node in the network being drained out of energy (as adopted in this chapter, since each sensor collects critical information in the network). Network lifetime is defined as the time duration until some target area is uncovered by any sensor node in [11,12]. In [13], it is defined as the time duration until the first occasion that data collection fails. For a detailed survey on the definition of network lifetime, refer to [14].

There are various techniques to save energy consumption and to maximize the network lifetime in WSNs (the definition of network lifetime in each work may be different though). In [15], the authors focused on the placement issues of wireless sensor nodes to increase the power efficiency. In [16], a data compression algorithm is proposed to reduce the amount of data to be transmitted for each sensor node, thus reducing the energy consumption. In the work of [17] and [18], the authors considered the routing problems to reduce energy consumption for wireless sensor nodes. A sensor node control approach that schedules the nodes' sleep/wakeup activities is studied in [19–22]. Among these aforementioned approaches, cross-layer design of the sensor network is an active research area [23].

Early studies of cross-layer design of WSNs focused on minimizing the total energy consumption. For example, in [24], the authors investigated the energy consumption minimization problem for an interference-free TDMA-based WSN through joint design of the PHY, MAC, and routing layers. By solving the approximated convex optimization problems, the results reveal that the minimum energy transmission scheme is a combination of multihop and single-hop transmissions. In [25], the energy consumption minimization problem is extended to a clustered WSNs where slot reuse and packet retransmission are considered to increase the network throughput.

However, it is pointed out that [26] the minimization of energy consumption may lead to some nodes being drained of energy rather rapidly. Therefore, the spatial information collected from the particular sensor node may be lost and this influences the following data analysis. Hence, instead of minimizing energy consumption of the sensor nodes, the authors in [26] attempted to maximize the network lifetime of the sensor network, which is defined as the first sensor node in the network being exhausted of energy. In [27], the joint design of the PHY, MAC, and the network layer is considered together with the transmission success probability to maximize the network lifetime. More recently, the authors in [28] considered the problem of network lifetime maximization with MAC-aware routing that is capable of multichannel access. In [29], contention and sleep control probabilities of each node are integrated into the network lifetime maximization problem. A joint design on the PHY, MAC and network layers to maximize the network lifetime considering spatially periodic time sharing scheme is proposed in [8] for a string topology. It is later extended to a fully connected WSN with random topology by introducing the concept of route lifetime for each node in [10].

The aforementioned work has considered different approaches to either minimize the energy consumption of sensor nodes or to maximize the network lifetime in a pure WSN. In this chapter, a hybrid sensor network is proposed. With the integration of both wireless sensor nodes and the PL nodes, the hybrid sensor network is expected to prolong network lifetime significantly.

9.4 System Model

A cluster-based sensor network for industrial automation system divides the industrial area into multiple clusters and the operation within each cluster is independent. As shown in Figure 9.1, each cluster covers a certain region and contains the sensors and a cluster head (or wired access point [4]), where critical messages such as the measurement and actuation information are collected by sensors and forwarded to the cluster head. The cluster heads will forward the data to the central controller by dedicated links (such as Ethernet) for further analysis.

A hybrid sensor network that includes both the wireless nodes and PL nodes (shown in Figure 9.2) is considered, which could be viewed as the sensor network in one of the clusters. The wireless nodes can be used to collect critical messages such as the measurement and actuation information. For example, the wireless nodes can be scatted in the chemical reaction pool to monitor the

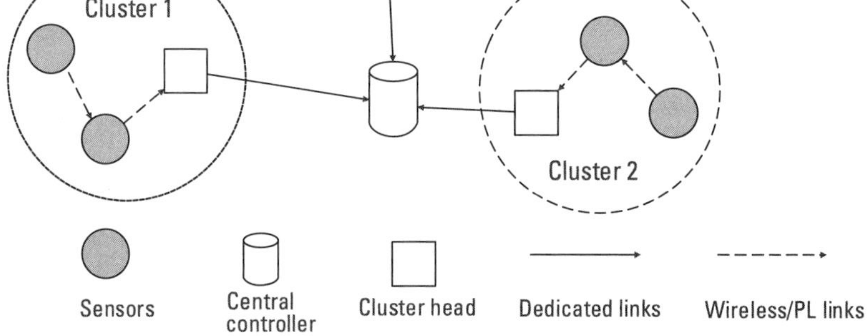

Figure 9.1 A clustered sensor network.

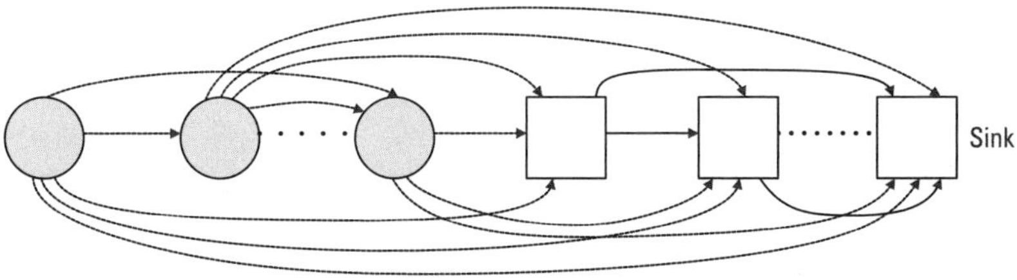

Figure 9.2 Topology of the hybrid sensor network, which could be viewed as the sensor network in one of the clusters. (The circles denote wireless nodes, the squares are PL nodes, the dashed arrows represent wireless links, and the solid arrows are PL links. The sink node is also a PL node).

process constantly. The PL nodes can be installed on the overhead power sockets on the ceiling and be used as complements of wireless nodes to collect critical information and relay data through the PL nodes (or be used as the wired network in the automation and management levels as introduced in the beginning of this chapter), so that the network lifetime can be prolonged.

It is assumed that the N sensor nodes in the hybrid sensor network are positioned in a line such that two neighboring nodes have a separation of d, and the first a nodes are wireless nodes, followed by $(N - a)$ PL nodes, as shown in Figure 9.2. The Nth node is considered as the sink (or cluster head, also a PL node) to which all the other nodes in the cluster will forward their collected data. It is also assumed that all the PL nodes are equipped with antennas to receive wireless signals from the wireless nodes (the received signals are downconverted to baseband by a lowpass filter, modulated by the PL carrier frequency, and then forwarded to the adjacent PL node. A prototype of such a platform can be found in [30]), so that the following transmission links exist;

- Wireless node → wireless node (through wireless link)
- Wireless node → PL node (through wireless link)
- PL node → PL node (through PL link)

Notations: The wireless nodes set is denoted by $\mathbb{W} = \{1, ..., a\}$ and the PL nodes set is denoted by $\mathbb{P} = \{a+1, ..., N\}$. L_w is defined as the number of wireless links and L_p is the number of PL links. $\mathcal{L}_w$ and $\mathcal{L}_p$ are used to label all the wireless and PL links, such that $\mathcal{L}_w = \{1, ..., L_w\}$ and $\mathcal{L}_p = \{1, ..., L_p\}$. $l_w \in \mathcal{L}_w$ is the index used to denote the lth wireless link and $l_p \in \mathcal{L}_p$ is used to represent the lth PL link. $\mathcal{L} = \mathcal{L}_w \bigcup \mathcal{L}_p$ denotes all the links in the hybrid sensor network and $l \in \mathcal{L}$ indicates the lth link, which is from a transmitter node i to a receiver node j and is denoted by (i, j). Finally, $\mathcal{O}(i)$ and $\mathcal{I}(i)$ represent the set of outgoing and incoming links at node i, respectively. x and $x^{'}$ are used to denote variables and parameters related to wireless and PL links, respectively.

9.4.1 Physical Layer

The path loss for the lth wireless link, G_{l_w}, is assumed to be given by $G_{l_w} = G_0/d_{l_w}^m$, where G_0 is the path loss at $d = 1$ m, d_{l_w} is the transmission distance for link l_w, and m is the path loss exponent with typical value $2 \leq m \leq 6$. The noise is assumed to be AWGN with single-sided power spectral density N_0. Let p_{l_w} denote the transmission power over the link l_w, then the SNR is,

$$\gamma_{l_w} = \frac{p_{l_w} G_{l_w}}{N_0 B} \tag{9.1}$$

where B is the transmission bandwidth for wireless links.

For PL nodes, NB PLC in band B (95–125 kHz), which is specified by CENELEC, is considered [31]. A deterministic propagation model is assumed for the PL channel, such that the power gain for PL link is given by

$$G_{l_p} = 10^{-a(f) \cdot d_{l_p}} \tag{9.2}$$

where f is the frequency, $a(f)$ varies between 0.004 m^{-1} (best case) and 0.01 m^{-1} (worst case), and d_{l_p} is the transmission distance for link l_p [32]. As in [31], the single-sided noise power spectral density is assumed to be $N_0^{'}$. Let p_{l_p} denote the transmission power, then the SNR over link l_p is given by

$$\gamma_{l_p} = \frac{p_{l_p} G_{l_p}}{N_0^{'} B^{'}} \tag{9.3}$$

where $B^{'}$ is the transmission bandwidth for PL links.

Therefore, for a general transmission link l, it follows that

$$\gamma_l = \begin{cases} \gamma_{l_w} & : \quad l \in \mathcal{L}_w \\ \gamma_{l_p} & : \quad l \in \mathcal{L}_p \end{cases} \tag{9.4}$$

M-ary quadrature amplitude modulation (MQAM) is assumed to be used in the sensor network. The data rate over link l is expressed as [24],

$$r_l = B_l \log_2(1 + K\gamma_l) \tag{9.5}$$

where B_l is the transmission bandwidth for link l and $K = -1.5/\ln(5BER)$ is the maximum possible coding gain with target BER, BER, for modulation schemes such as MQAM [33]. This data rate is upper bounded by the maximum allowable transmission power, p_{max}, for wireless links and the PSD mask, $\bar{p}$, for PL links. The maximum data rate can be achieved for wireless and PL links (i.e., C_{l_w} and C_{l_p}) can be obtained by substituting p_{max} and $\bar{p}$ into (9.4) and (9.5), respectively. Therefore, the maximum transmission rate constraint is

$$r_l \leq C_l \tag{9.6}$$

where

$$C_l = \begin{cases} C_{l_w} & : \quad l \in \mathcal{L}_w \\ C_{l_p} & : \quad l \in \mathcal{L}_p \end{cases} \tag{9.7}$$

9.4.2 MAC Layer

The TDMA approach is considered as the MAC scheme for both transmission mediums due to interference free and energy saving advantages. In a slotted synchronous TDMA MAC scheme, each frame of length T is divided into multiple slots of length Δ. It is assumed that only one wireless/ PL link is allowed to transmit in each time slot in the wireless/ PL medium. Also note that wireless links and PL links follow a TDMA approach separately. Therefore, within each TDMA frame of length T, if link l is allocated n_l slots, it transmits for time

$$t_l = n_l \Delta \tag{9.8}$$

Note that the transmission can occur in the wireless link and PL link simultaneously and the interference between wireless links and PL links is neglected (e.g., due to the coupling circuit in the PL nodes) since they operate in two different frequency bands. Therefore, it follows that

$$\begin{cases} \sum_{l \in \mathcal{L}_w} t_l & \leq \quad T \\ \sum_{l \in \mathcal{L}_p} t_l & \leq \quad T \end{cases} \tag{9.9}$$

This means the transmission in either wireless or PL links should be completed within time period T.

9.4.3 Traffic Flow

A single commodity flow [24] is assumed in this work where each node collects energy management information, such as temperature and humidity, which needs to be forwarded to a single sink node. Assuming each node (except the sink node) collects data at the same rate of R, then the Sink node (Nth node) will have data rate

$$R_N = -\sum_{i=1}^{N-1} R_i \tag{9.10}$$

where the negative sign indicates that the sink node only has incoming traffic. If data is transmitted over link l with a data rate of r_l for time t_l, then the amount of data transmitted over link l in period T is

$$W_l = r_l t_l \tag{9.11}$$

In particular, the flow conservation constraints should be satisfied at the end of every time period T; that is, the difference between the outgoing data and incoming data is equal to the data collected locally of each node i,

$$\sum_{l \in \mathcal{O}(i)} W_l - \sum_{l \in \mathcal{I}(i)} W_l = R_i T \tag{9.12}$$

9.5 Problem Formulation

In order to investigate the hybrid network lifetime, which is defined as the time duration until the first wireless node being drained of energy, it is assumed that the PL nodes are powered by PL and thus have unlimited power supply, while the wireless nodes are powered by nonrechargeable batteries (wireless nodes are likely to be placed where mains power cannot be attached, for example the wireless nodes scatted in the chemical reaction pool and can not be attached to the mains power due to safety concerns). Therefore, this work only focuses on the energy consumption model for the wireless nodes. As well, only the energy consumption for transmission is considered to illustrate the main ideas (as widely adopted in, for example [8,10,26]). Therefore, if the transmission power over wireless link l_w is p_{l_w}, the power consumption at the transceiver circuit of wireless transmitter node i of link l_w is given by

$$P_{transceiver} = (1 + \alpha) p_{l_w} \tag{9.13}$$

where α is the inefficiency of the power amplifier and is taken to be a constant [34].

The actual power consumption of the sensor should take into account the power consumption for the active mode, the sleep mode, and the transient mode [35]. The active mode power includes the transmission signal power and the circuit power consumption. For the ease of derivation, and as widely adopted in [8,10,26], only the power consumption of transmission signal power and the power consumption of the power amplifier is considered. In the sleep mode the power consumption is dominated by the leakage current of the switching transistors. Since for analog circuits the leakage power consumption is usually much smaller than the power consumption in the active mode, leakage power is neglected in the total energy consumption. Also, since the duration of the transient mode is much smaller as compared to the slot time, the power consumption for the transient mode is relatively small and not considered in this chapter. Some notations of the parameters used in the problem formulation are summarized in Table 9.1.

From the wireless communication model, the power consumption required to support the data rate, $\frac{W_{l_w}}{t_{l_w}}$, over link l_w is given by

$$p_{l_w} = \frac{N_0 B}{K G_{l_w}} \left(2^{\frac{W_{l_w}}{B t_{l_w}}} - 1\right) \tag{9.14}$$

Table 9.1
Parameters Used in Problem Formulation

T_{net}	Network lifetime
q	The inverse of T_{net}
E_c	Initial energy
R_i	Data arrival rate for the ith node
T	Time frame
Δ	Time slot duration
a	Number of the wireless sensor node
N	Number of the total sensor node
α	Power inefficiency
$p_{i,j}$	Transmission power over wireless link $i \rightarrow j$
$W_{i,j}$, $W'_{i,j}$	Data transmitted over wireless, PL link $i \rightarrow j$ in period T
$t_{i,j}$, $t'_{i,j}$	Transmission time allocated to wireless, PL link $i \rightarrow j$ in period T
$C_{i,j}$, $C'_{i,j}$	Maximum transmission data rate over wireless, PL link $i \rightarrow j$

This is derived as follows. By substituting (9.1) into (9.5), it has

$$r_{l_w} = B \log_2 \left(1 + K \frac{p_{l_w} G_{l_w}}{N_0 B}\right) \tag{9.15}$$

Note that r_{l_w} here is the data rate on the wireless link (notation l_w instead of l), and can be represented by $\frac{W_{l_w}}{t_{l_w}}$, and therefore, it has

$$\frac{W_{l_w}}{t_{l_w}} = B \log_2 \left(1 + K \frac{p_{l_w} G_{l_w}}{N_0 B}\right) \tag{9.16}$$

or

$$2^{\frac{W_{l_w}}{B t_{l_w}}} = 1 + K \frac{p_{l_w} G_{l_w}}{N_0 B} \tag{9.17}$$

From the above equation, (9.14) can be obtained.

Therefore, assuming the initial battery energy is E_c, the lifetime of wireless node i is defined as

$$T_i = \frac{E_c}{\Sigma_{l_w \in \mathcal{O}(i)} p_{l_w} (1 + \alpha) \frac{t_{l_w}}{T}} \tag{9.18}$$

where $i \in \mathbb{W}$. Then the network lifetime can be represented as

$$T_{net} = \min_{i \in \mathbb{W}} T_i \tag{9.19}$$

The objective is to maximize the hybrid network lifetime. Note that the original objective function T_{net} is not convex; however, it can be transformed into a convex function by taking an inverse of the network lifetime, $1/T_{net}$. It can be easily proved that the objective function after transformation, $1/T_{net}$, is jointly convex over W_l and t_l.

Therefore, the network lifetime maximization problem can be formulated as a mixed integer convex optimization problem [36]:

$$
\begin{array}{llll}
 & \min & 1/T_{net} & \\
s.t. & T_{net} & \leq & E_c/\Sigma_{l\in\mathcal{O}(i)} p_l(1+\alpha)\frac{t_l}{T} \quad (i\in\mathbb{W}) \\
 & C_l\cdot t_l & \geq & W_l \\
 & W_l & \geq & 0 \\
 & t_l & \in & \{0, \Delta, 2\Delta, ...\}
\end{array}
\tag{9.20}
$$

along with (9.9) and (9.12).

The variables to determine are T_{net}, W_l and t_l. The first inequality in (9.20) means that the energy consumption for each wireless node should not exceed the battery capacity for the whole network lifetime. The second inequality is the maximum transmission power constraint for all links. Constraint (9.9) ensures that the data relaying in either the wireless or PL link is finished within time period T. Equality (9.12) is the flow conservation constraints for all nodes. For the above problem formulation, if the integer constraints are relaxed on t_l, the optimization problem would be a convex problem.

9.6 Optimization Approach

In this section, the KKT conditions are used to derive analytical expressions for the hybrid network lifetime from the relaxed problem for the linear topology, which provide upper bounds on the network lifetime for the optimization problem in (9.20). The Lagrangian of the optimization problem in (9.20) is

$$
\begin{aligned}
\mathbf{L}(q, \mathbf{t}, \mathbf{W}, \lambda, \mu, \varphi, \varphi', \nu) = & \sum_{i\in\mathbb{W}} \lambda_i \cdot \left[\sum_{l\in\mathcal{O}(i)} \beta_{l_w} \cdot (2^{\frac{W_{l_w}}{B\cdot t_{l_w}}} - 1) \cdot t_{l_w} - E_c \cdot q \cdot T \right] \\
& + \sum_{l\in\mathcal{L}} \mu_l \cdot (W_l - C_l \cdot t_l) \\
& + \varphi \cdot (\sum_{l\in l_w} t_l - T) + \varphi' \cdot (\sum_{l\in l_p} t_l - T) \\
& + \sum_{i\in\mathbb{W}\bigcup\mathbb{P}} \nu_i \cdot (\sum_{l\in\mathcal{O}(i)} W_l - \sum_{l\in\mathcal{I}(i)} W_l - R_i T) \\
& + q
\end{aligned}
\tag{9.21}
$$

where $q = 1/T_{net}$ and $\beta_l = \frac{BN_0(1+\alpha)}{G_l K}$. λ and μ are the Lagrange multipliers associated with the first and the second inequality constraints in (9.20), respectively. φ and φ' are the Lagrange multipliers associated with (9.9) for the wireless and PL links, respectively. ν is the Lagrange multiplier with respect to (9.12).

The KKT conditions [36] are

$$1 - E_c T(\lambda_1 + \lambda_2 + \cdots + \lambda_a) = 0 \quad (9.22)$$

$$\frac{\beta_{1,2}}{B} \cdot 2^{\frac{W_{1,2}}{Bt_{1,2}}} \cdot \lambda_1 + \mu_{1,2} + \nu_1 - \nu_2 = 0 \quad (9.23)$$

$$\cdots$$

$$\frac{\beta_{1,a+1}}{B} \cdot 2^{\frac{W_{1,a+1}}{Bt_{1,a+1}}} \cdot \lambda_1 + \mu_{1,a+1} + \nu_1 - \nu_{a+1} = 0 \quad (9.24)$$

$$\cdots$$

$$\frac{\beta_{a,a+1}}{B} \cdot 2^{\frac{W_{a,a+1}}{Bt_{a,a+1}}} \cdot \lambda_a + \mu_{a,a+1} + \nu_a - \nu_{a+1} = 0 \quad (9.25)$$

$$\beta_{1,2}[(1 - \frac{W_{1,2}}{Bt_{1,2}}) \cdot 2^{\frac{W_{1,2}}{Bt_{1,2}}} - 1] \cdot \lambda_1 - C_{1,2} \cdot \mu_{1,2} + \varphi = 0 \quad (9.26)$$

$$\cdots$$

$$\beta_{a,a+1}[(1 - \frac{W_{a,a+1}}{Bt_{a,a+1}}) \cdot 2^{\frac{W_{a,a+1}}{Bt_{a,a+1}}} - 1] \cdot \lambda_a - C_{a,a+1} \cdot \mu_{a,a+1} + \varphi = 0 \quad (9.27)$$

$$\mu_{a+1,a+2} + \nu_{a+1} - \nu_{a+2} = 0 \quad (9.28)$$

$$\mu_{N-1,N} + \nu_{N-1} - \nu_N = 0 \quad (9.29)$$

$$-C'_{a+1,a+2} \cdot \mu_{a+1,a+2} + \varphi' = 0 \quad (9.30)$$

$$\cdots$$

$$-C'_{N-1,N} \cdot \mu_{N-1,N} + \varphi' = 0 \quad (9.31)$$

In particular, the complementary slackness conditions are

$$\lambda_1 \cdot [\beta_{1,2} \cdot (2^{\frac{W_{1,2}}{Bt_{1,2}}} - 1) \cdot t_{1,2} + \ldots$$

$$+\beta_{1,a+1} \cdot (2^{\frac{W_{1,a+1}}{Bt_{1,a+1}}} - 1) \cdot t_{1,a+1} - E_c \cdot q \cdot T] = 0 \quad (9.32)$$

$$\cdots$$

$$\lambda_a \cdot [\beta_{a,a+1} \cdot (2^{\frac{W_{a,a+1}}{Bt_{a,a+1}}} - 1) \cdot t_{a,a+1} - E_c \cdot q \cdot T] = 0 \quad (9.33)$$

$$\mu_{1,2} \cdot (W_{1,2} - C_{1,2} \cdot t_{1,2}) = 0 \quad (9.34)$$

$$\cdots$$

$$\mu_{a,a+1} \cdot (W_{a,a+1} - C_{a,a+1} \cdot t_{a,a+1}) = 0 \quad (9.35)$$

$$\cdots$$

$$\mu_{N-1,N} \cdot (W_{N-1,N} - C_{N-1,N} \cdot t_{N-1,N}) = 0 \tag{9.36}$$

$$\varphi \cdot [(t_{1,2} + t_{1,a+1} + \cdots + t_{a,a+1}) - T] = 0 \tag{9.37}$$

$$\varphi' \cdot [(t_{a+1,a+2} + t_{a+1,N} + \cdots + t_{N-1,N})] = 0 \tag{9.38}$$

$$\lambda, \mu, \varphi, \varphi', \nu \geq 0 \tag{9.39}$$

In the linear topology, all the wireless sensor nodes collect data. In the network configuration shown in Figure 9.2, the network lifetime will be maximized if all the wireless nodes relay their data to the first PL node (the $(a+1)$-th node), since the transmission distance is shortest compared to other PL nodes. Therefore, in order to simplify the derivation, four terms (i.e., $W_{i,i+1}$, $W_{i,a+1}$, $W'_{i,i+1}$, and $W'_{i,N}$) are considered for optimization. This means all the wireless nodes only relay data to the next wireless node or to the first PL node and the PL nodes relay data to the next PL node or to the sink node directly.

Assuming that all the wireless links use the same constellation size in MQAM. Then it follows that

$$\frac{W_{1,2}}{t_{1,2}} = \frac{W_{1,a+1}}{t_{1,a+1}} = \frac{W_{2,3}}{t_{2,3}} = \frac{W_{2,a+1}}{t_{2,a+1}} = \cdots = \frac{W_{a,a+1}}{t_{a,a+1}} \tag{9.40}$$

Note that in an optimal optimization scheme, all the wireless nodes should be drained of energy at the same time. Therefore, it can be obtained from (9.32) and (9.33) that

$$\begin{aligned} \beta_{1,2}(2^{\frac{W_{1,2}}{Bt_{1,2}}} - 1) \cdot t_{1,2} + \beta_{1,a+1}(2^{\frac{W_{1,a+1}}{Bt_{1,a+1}}} - 1) \cdot t_{1,a+1} &= \cdots \\ &= \beta_{a,a+1}(2^{\frac{W_{a,a+1}}{Bt_{a,a+1}}} - 1) \cdot t_{a,a+1} \end{aligned} \tag{9.41}$$

By following a similar procedure in [37], and substituting (9.40) into (9.41), and by further derivation, yields

$$\begin{aligned} W_{1,2} + a^m W_{1,a+1} &= W_{2,3} + (a-1)^m W_{2,a+1} = \cdots \\ &= W_{a-1,a} + 2^m W_{a-1,a+1} = W_{a,a+1} \end{aligned} \tag{9.42}$$

Now from the flow conservation constraint (9.12), it follows that

$$W_{a,a+1} - W_{a-1,a} = RT \tag{9.43}$$

From (9.42), it has

$$W_{a-1,a} + 2^m W_{a-1,a+1} = W_{a,a+1} \tag{9.44}$$

Solving (9.43) and (9.44), gives

$$W_{a-1,a+1} = \frac{RT}{2^m} \tag{9.45}$$

Combining (9.12) and (9.42), and by repeating the process above, all the values of $W_{i,j}$ can be determined. Now with the values of $W_{i,j}$, the values of $t_{i,j}$ can be obtained from (9.40).

The optimal values of $W_{i,j}$, and $t_{i,j}$ are summarized as follows:

$$W_{i,a+1} = \begin{cases} \frac{RT+[(a-i)^m-1]W_{i+1,a+1}}{(a+1-i)^m} & : \quad 1 \leq i \leq a-2 \\ \frac{RT}{2^m} & : \quad i = a-1 \end{cases} \tag{9.46}$$

$$W_{i,i+1} = \begin{cases} iRT - \sum_{j=1}^{i} W_{j,a+1} & : \quad 1 \leq i \leq a-1 \\ aRT - \sum_{j=1}^{a-1} W_{j,a+1} & : \quad i = a \end{cases} \tag{9.47}$$

$$\begin{cases} t_{i,a+1} = \frac{W_{i,a+1}}{W_M+W_N} \cdot T \\ t_{i,i+1} = \frac{W_{i,i+1}}{W_M+W_N} \cdot T \end{cases} \tag{9.48}$$

where $W_M = \sum_{i=1}^{a-1} W_{i,a+1}$ is the total amount of data transmitted from each wireless node (except the ath node) to the 1-st PL node and $W_N = \sum_{i=1}^{a} W_{i,i+1}$ is the total amount of data transmitted from each wireless node to the next node.

Since the PL nodes have unlimited power supply, it is assumed that all the PL nodes will relay their data to the next PL node (the effectiveness of this transmission strategy will be examined in Section 9.7) and all PL nodes transmit at the maximum power, and therefore, $W'_{i,j}$ and $t'_{i,j}$ can be calculated as

$$\begin{cases} W'_{i,i+1} = iRT & : \quad a+1 \leq i \leq N-1 \\ t'_{i,i+1} = \frac{W'_{i,i+1}}{C'_{i,i+1}} & : \quad a+1 \leq i \leq N-1 \end{cases} \tag{9.49}$$

From (9.33), and the results of $W_{a,a+1}$ and $t_{a,a+1}$, the closed-form expression for the hybrid network lifetime maximization can be derived as

$$T_{net} = \frac{E_c(W_M + W_N)}{\beta_{a,a+1}(aRT - W_M)(2^{\frac{W_M+W_N}{BT}} - 1)} \tag{9.50}$$

Equation (9.50) provides the globally optimal value for hybrid network lifetime maximization. Since the integer constraints on t_l are relaxed, the closed-form expression of network lifetime maximization in (9.50) provides an upper bound to the network lifetime of the hybrid sensor network.

In (9.50), since $\beta_{a,a+1}$ is a constant, it is clear that the network lifetime of the hybrid network in the linear topology is only related to the initial battery capacity, E_c, the number of battery powered wireless nodes a, and the data arrival rate R. While the closed-form expression in the linear topology is a complication, it facilities to sensor network system design by fast parameter fitting to investigate the relationship between the network lifetime and the related system parameters. As well, the optimal routing strategy in the linear topology is multipath multihop routing.

9.7 Numerical Results and Analysis

In this section, numerical results obtained from the proposed approach are used to demonstrate the effectiveness of the proposed hybrid sensor network. The parameters used are summarized in Table 9.2. The radio frequency operates at 2.4 GHz in the industrial-scientific-medical (ISM) band, with a reference path loss at $d = 1m$ of 40 dB. The initial energy of the wireless node is 5,000J, which is typical for the capacity of an AAA alkaline long-life battery [8]. The maximum transmit power of the wireless node is 10 mW, as specified in the standard in [38], and is used as in the literature in [39].

To test the accuracy of the derived closed-form network lifetime expression, the separation between adjacent nodes is set to 1.5m, and the total number of sensors is 10 with 5 PL nodes. Figure 9.3 depicts the comparison of network lifetime obtained from the closed-form expression with the results obtained by CVX [40] and OMNeT++ [41,42]. The results shown that the network lifetime obtained from the closed-form expression is slightly longer than the

Table 9.2
Parameters Used for Simulation

Symbol	Value	Description
f_c	2.4 GHz	Radio frquency
E_c	5000 J [8]	Initial energy of wireless node
p_{max}	10 mW [24]	Maximum Tx power of wireless node
N_0	−105 dBm/Hz	Noise PSD level of wireless link
B	10 kHz	Bandwidth of wireless link
d	1.5 m	Distance between adjacent nodes
T	1 s	Time frame
Δ	0.001T	Time slot duration
G_0	40 dB	Path loss at $d = 1$ m
m	3.5	Path loss exponent
α	1.9	Power inefficiency
BER	10^{-3}	Target BER
f'_c	110 kHz	Carrier frequency of PL link
B'	30 kHz	Bandwidth of PL link
N'_0	−80 dBm/Hz	Noise PSD level of PL link
$\bar{p}$	−25 dBm/Hz	PSD mask

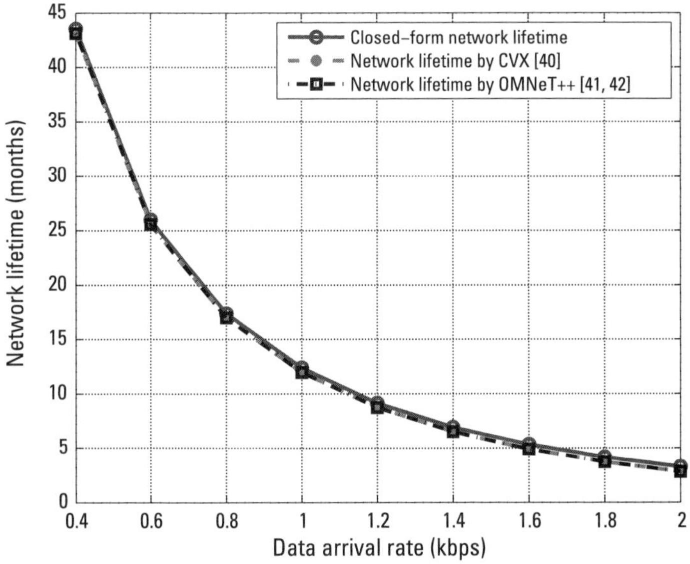

Figure 9.3 Comparison of network lifetime in linear topology with CVX [40] and OMNeT++ [41,42].

results obtained by the other two methods, especially when the data arrival rate is relatively larger. This is due to the fact that in deriving the closed-form network lifetime expression, the integer constraints on t_l are relaxed. Therefore, it provides an upper bound to the network lifetime. However, as shown in Figure 9.3, this upper bound is very tight to the optimal solution when integer constraints are taken into account.

In Figure 9.4, the network lifetime of a hybrid sensor network is compared with a pure WSN (when $a = N$) with the same total number of sensors (including the sink node). Under different data arrival rates, the number of PL nodes in the hybrid sensor network is fixed to five regardless of total number of nodes. The separation between adjacent nodes is set to 1.5m. It is obvious that the lifetime of the hybrid sensor network is much longer than that of the WSN, especially when the network scale is relatively small and the data arrival rate is relatively low. For example, with a total of 12 sensors and a data arrival rate of 1 kbps, the hybrid network lifetime is enhanced by eight times compared to a pure WSN with the same total number of sensors and with the same data arrival rate. This advantage is due to the fact that all the wireless nodes forward their data to the first PL node in the hybrid network, thus reducing transmission distances for wireless links. Also, since wireless links and PL links use two different mediums for transmission, each wireless node in a hybrid network can access a wireless channel for a longer time (the PL nodes use PL channel for transmission and does not interfere with the wireless channel), thus reducing the power requirement for

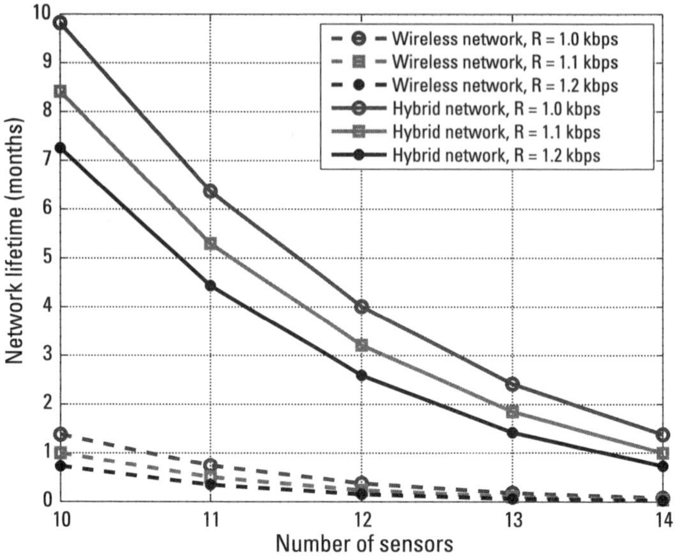

Figure 9.4 Network lifetime versus the total number of sensor nodes in linear topology (with five PL sensors for the hybrid network).

each wireless link. However, when the number of sensors is large, the lifetime of both networks drops rapidly due to lack of enough energy.

Figure 9.4 also shows that with the increase of total number of sensors, the lifetime of both networks decreases. This is mainly due to the fact that with an increase in the total number of sensors, the network scale increases and consequently the transmission distances of wireless nodes to the sink (in WSN) and to the first PL node (in the hybrid sensor network) increases, thus the wireless links suffer from poorer channel conditions. Also, since the time period T is fixed, each node is allocated less transmission time with the increase of the total number of sensors. Therefore, with the same data arrival rate, each sensor node tends to use more power for transmission and causes a reduction in the network lifetime. Each sensor node will also consume more energy for transmission with the increase of data arrival rate. This explains the reduction of the network lifetime with the increase of the data arrival rate.

In the linear topology, the multihop transmission mechanism is used, and the joint cross-layer optimization results show that the optimal solution of time allocation on each link is not a uniform TDMA. Therefore, it is necessary to compare the network lifetime of the hybrid sensor network using the cross-layer optimization method and single-layer optimization method. Figure 9.5 compares the network lifetime of a hybrid sensor network using the cross-layer optimization method and single-layer optimization method (where a uniform TDMA mechanism is adopted for link time allocation, and the amount of data

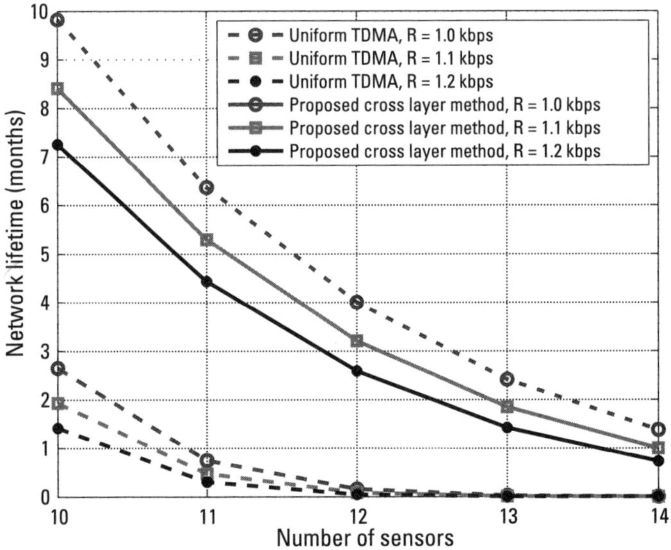

Figure 9.5 Network lifetime versus the total number of sensor nodes in linear topology (with five PL sensors for the hybrid network).

to be transmitted on each link is optimized, labeled as "Uniform TDMA" in Figure 9.5) under different total number of sensor nodes and under different data arrival rates. The network setup is exactly the same as the network setup of the hybrid sensor network in Figure 9.4. Figure 9.5 depicts that the network lifetime of the hybrid sensor network adopting cross-layer optimization method is much longer than that of the single-layer optimization method, especially when the network scale is relatively small and the data arrival rate is relatively low. This proves the advantage of the cross-layer optimization method on the linear topology.

Since the linear network topology considered in this chapter can be viewed as a cluster in a sensor network for an industrial automation system, the number of nodes in each cluster could be small, while the span of the network matters. Therefore, as shown in Figure 9.6, the network lifetime performance of hybrid networks is compared under different sensor densities. The hybrid network with a total of 16 nodes including five PL nodes as in Figure 9.4 is used as a reference for comparison. The total length of the network is 22.5m (15 segments multiplied by the separation of 1.5m). With a total of 16 nodes, the separation between two adjacent nodes is 1.5m, while the separation is 2.5m with 10 nodes. The rest simulation parameters are the same as in Table 9.2. From Figure 9.6, the hybrid network with lower sensor density outperforms the reference network in terms of the network lifetime, especially when the data arrival rate is low. This is due to the fact that although the distance between adjacent nodes are larger in the hybrid

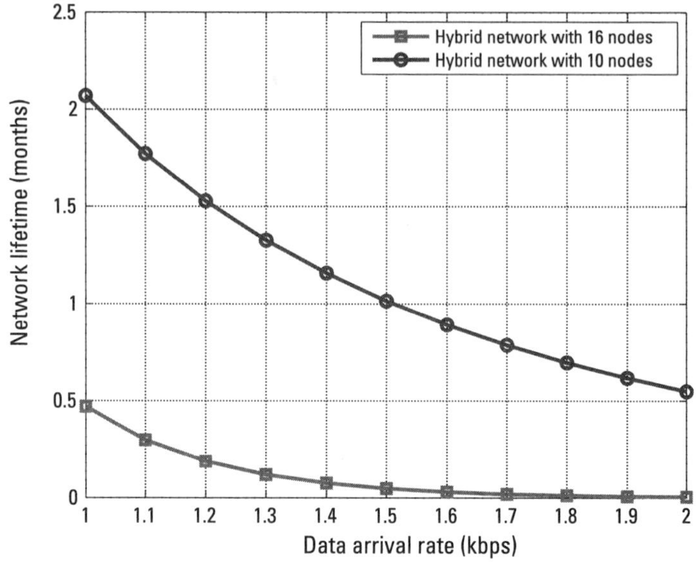

Figure 9.6 Impact of the data arrival rate on the network lifetime with a fixed total network range of 22.5 m.

network with lower sensor density, the distance between each wireless node to the first PL node is reduced. Also, for a hybrid network with lower sensor density, each wireless node can access the wireless medium for a longer time, and thus can reduce the transmission power.

However, one may argue that the network lifetime is prolonged since the total amount of data forwarded to the sink is decreased, which is true because with same data arrival rate, 15 nodes will collect data in the reference network while only nine nodes collect data in the hybrid network for comparison. It can be noticed from Figure 9.6 that even with the same total amount of data collected, the hybrid network with a lower sensor density still exhibits a much longer lifetime than the reference one. For example, when the data arrival rate for the reference network is 1 kbps, the total amount of data collected in time period T is 15kb, and the corresponding lifetime is around 0.5 month. This means each node in the hybrid network with lower sensor density should collect data at a rate of 1.67 kbps to maintain the same amount of data collected. While at $R = 1.67$ kbps, the corresponding lifetime is around 0.8 month, which is still much longer than that of the reference network. Note that since a bandwidth of 10 kHz is considered, the data arrival rate in this chapter is generally low, which could be suitable for scalar information (such as temperature and humidity) collection.

Note that in Figure 9.6, when the data rate is increased from 1 kbps to 2 kbps, the network lifetime of the hybrid network with 16 nodes drops significantly. This is due to the fact that only the power consumption of the active mode of the sensor network is considered. Also, for the ease of derivation of the closed-form expression, only the power consumption of transmission signal power and the power consumption of the power amplifier is considered as widely adopted in the literature [8,10,26]. By considering the complete power consumption model would make the derivation very difficult and is left as a future work. In this case, another reason is that since the hybrid network has 16 nodes, each wireless node performs data collection. When the data arrival rate is increased from 1 kbps to 2 kbps, the overall amount of data collected by the sensor network becomes large and the sensor network may not be able to accommodate the required amount of traffic and hence results in a very low network lifetime.

Based on the above analysis, the hybrid sensor network has a significant improvement in prolonging the network lifetime compared to a pure WSN. Also, by limiting the data arrival rate of sensor nodes and the sensor density in a linear hybrid sensor network with given length and fixed PL nodes, the hybrid network lifetime can be maximized.

9.7.1 Transmission Strategies of PL Nodes

In order to investigate the transmission strategies of PL nodes in the CENELEC band B (95–125 kHz) for a linear hybrid sensor network (i.e., each PL node transmit data to the next PL node or to the sink directly), the PL network topology is constructed as shown in Figure 9.7 (the rightmost PL node is the sink), where five PL nodes (labeled as node 1 to 5 in Figure 9.7) are placed on the main span. These PL nodes are arranged so that they are evenly spaced at a separation of 1.5m. The topology of the branches attached to each PL node is randomly generated according to the rules described as follows. Each branch attached to node 1 to 5 in Figure 9.7 may have 5 to 10 outlets with bus or star connection and the distances between adjacent outlets (or between the outlets to the PL nodes) varies from 1m to 5m. As well, it is assumed that each outlet is randomly connected with an appliance that exhibit impedances of 5Ω, 50Ω, 150Ω, 1,000Ω, or frequency-selective impedances [43] with resistance at resonance equal to 394Ω, 863Ω, or

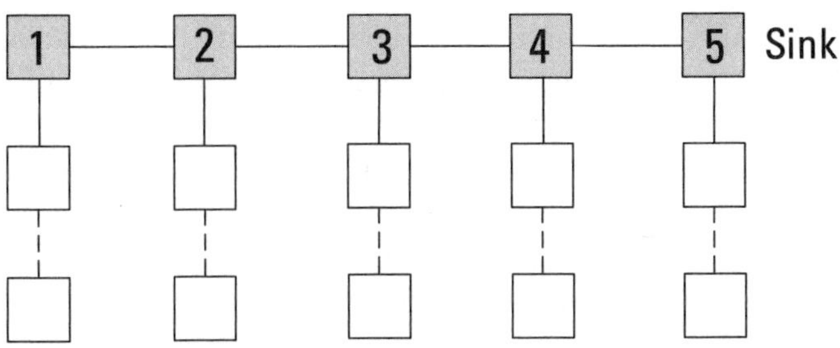

Figure 9.7 PL network topology (the gray squares represent the PL nodes in the hybrid sensor network while the white squares denote the outlets with bus or star connection).

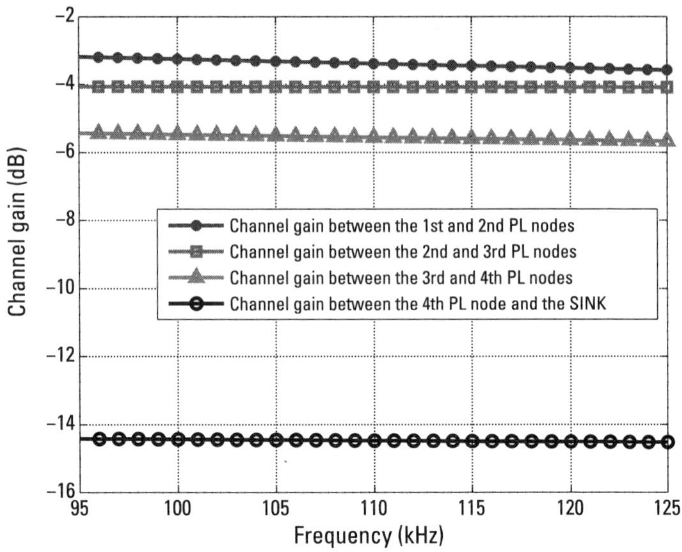

Figure 9.8 Average channel gain between two adjacent PL nodes.

1312Ω. Please refer to [43] for the frequency-selective impedance model and the relative parameters. In addition, the impedance of PL nodes are assumed to be 100Ω and transmission cable type 1 in [43] is considered in this chapter. For the simulation, the impedance carry-back method and VRA in [44] is utilized to determine the channel frequency responses between each adjacent PL node pairs and between each PL node to the Sink. The frequency responses for an example PL network topology are shown in Figures 9.8 and 9.9, respectively. Note that the PL node farthest to the sink is denoted as the 1st PL node.

Comparing Figures 9.8 and 9.9, noted that the PL channel between each PL node and the next PL node toward the sink, in general, has a better channel performance than that between each PL node and the Sink node directly. This is mainly due to the fact that without intermediate PL nodes, the PL channel suffers from less path loss due to the multipath propagation effect and thus exhibits better channel performance. This is also valid when the PL nodes are placed at a larger distance (e.g., 15m) and thus provides larger network coverage. This fact supports the proposed transmission strategies that each PL node should relay their data to the PL node next to it. To simulate the achievable data rate for the PL links, the parameters shown in Table 9.2 are used. From extensive simulations, the maximum achievable data rate for the transmission of each PL node to the next PL node is between 100 kbps and 500 kbps depending on the channel conditions. Such a maximum achievable data rate is sufficient to support the data relaying for a small-scale hybrid sensor network with low data arrival rate within given time period T.

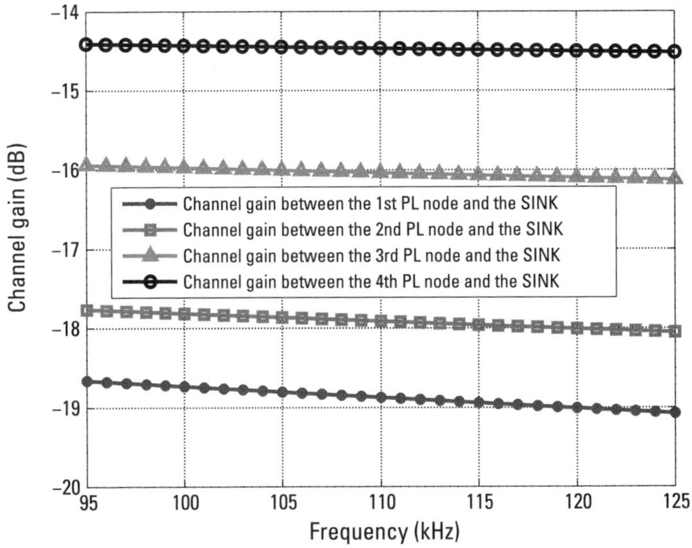

Figure 9.9 Average channel gain between each PL node and the Sink.

9.8 Summary

In this chapter, a cross-layer design of a hybrid sensor network was proposed to maximize the network lifetime. The proposed system is suitable in situations where the wireless signal may not penetrate or in situations where wired communication infrastructure is a problem (e.g., installing new communication systems in old facilities or the cost of installing wires for communication within refineries is very high due to safety requirements). As well, the optimal transmission scheme is obtained and the closed-form expression for the globally optimal solution for network lifetime maximization is derived for the linear network topology. From the closed-form expressions, it is clear that the network lifetime of the hybrid network is only related to the initial battery capacity, the number of battery-powered wireless nodes, and the data arrival rate. While the closed-form expression in the linear topology is complicated, it facilities sensor network system design by fast parameter fitting to investigate the relationship between the network lifetime and the related system parameters. It is also obvious that the optimal routing strategy in the linear topology is multipath multihop routing. The proposed cross-layer optimization method outperforms the single-layer optimization method since multipath multihop routing in linear topology allows more design freedom.

The results show that in the linear topology, the hybrid sensor network enables a significant increase in the network lifetime over the pure WSN, especially with a relatively low data arrival rate. For example, in the linear topology, with five PL nodes in the hybrid network with a total of 12 sensors and a data arrival rate of 1 kbps, the network lifetime is prolonged by eight times compared to a pure WSN with the same total number of sensors and with the same data arrival rate. Simulation results also validate the advantage of cross-layer optimization method over individual-layer optimization method on the linear topology. In addition, the simulation results reveal that the hybrid sensor network with a lower sensor node density exhibits a much longer network lifetime even when the same amount of data is collected in the linear topology. It can be concluded that closed-form expressions provide a useful guideline for the design of the sensor network. As well, the hybrid sensor network prolongs network lifetime mainly by a combination of increasing channel access time and reducing transmission distance of the wireless link. By carefully setting the distance separation, data arrival rate, and the sensor density in a hybrid network with given number of PL nodes, the network lifetime can be maximized.

The network lifetime of the hybrid network in the linear topology is compared to a pure battery-powered WSN with the same network configurations in this chapter, and the hybrid network has a much longer network lifetime. However, intuitively, in the pure wireless network with the same number of total sensors as compared to the hybrid network, assuming that the wireless nodes that are placed at the same positions of the PL nodes in the corresponding hybrid network are mains powered. In this case, in the hybrid network, the transmission in the wireless link and in the PL link can occur concurrently. While in the wireless network, although the wireless nodes at the same positions of the PL nodes in the hybrid network are also mains-powered, the transmission of these wireless nodes will interfere with the transmission of the wireless links of the battery powered wireless nodes. Since TDMA is considered in this chapter and concurrent transmission is not allowed in the same transmission medium, the transmission of the mains-powered wireless nodes in the wireless network will have to share the transmission time frame with the battery-powered wireless nodes. Therefore, each battery-powered wireless node in the wireless network is allocated less transmission time as compared to the battery-powered wireless node in the hybrid network, which consequently consumes more transmission power in order to achieve the required data arrival rate. While this is currently too complicated to be analyzed mathematically, and algorithms can be developed to solve the problem in this situation, this is left as a future work.

References

[1] Goldsmith, A., *Wireless Communications*, Cambridge, United Kingdom: Cambridge University Press, 2005.

[2] Gaj, P., J. Jasperneite, and M. Felser, "Computer Communication within Industrial Distributed Environment: A Survey," *IEEE Transactions on Industrial Informatics*, Vol. 9 No. 1, 2013, pp. 182–189.

[3] Zoppi, S., A. Van Bemten, H. Murat Gursu, M. Vilgelm, J. Guck, and W. Kellerer, "Achieving Hybrid Wired/Wireless Industrial Networks with WDetServ: Reliability-Based Scheduling for Delay Guarantees," *IEEE Transactions on Industrial Informatics*, Vol. 14, No. 5, 2018, pp. 2307–2319.

[4] Herrmann, M. J., and G. G. Messier, "Cross-Layer Lifetime Optimization for Practical Industrial Wireless Networks: A Petroleum Refinery Case Study," *IEEE Transactions on Industrial Informatics*, Vol. 14, No. 8, 2018, pp. 3559–3566.

[5] Johnstone, I., J. Nicholson, B. Shehzad, and J. Slipp, "Experiences from a Wireless Sensor Network Deployment in a Petroleum Environment," in *Proceedings of the 2007 International Conference on Wireless Communications and Mobile Computing*, Honolulu, Hawaii, 2007 pp. 382–387.

[6] Pereira, S. C., A. S. Caporali, and I. R. S. Casella, "Power Line Communication Technology in Industrial Networks," in *2015 IEEE International Symposium on Power Line Communications and Its Applications (ISPLC)*, Austin, Texas, 2015, pp. 216–221.

[7] Jung, J. W., and M. A. Weitnauer, "On Using Cooperative Routing for Lifetime Optimization of Multi-Hop Wireless Sensor Networks: Analysis and Guidelines," *IEEE Transactions on Communications*, Vol. 61, No. 8, 2013 pp. 3413–3423.

[8] Yetgin, H., K. T. K. Cheung, M. El-Hajjar, and L. Hanzo, "Cross-Layer Network Lifetime Maximization in Interference-Limited WSNs," *IEEE Transactions on Vehicular Technology*, Vol. 64, No. 8, 2015, pp. 3795–3803.

[9] Cassandras, C. G., T. Wang, and S. Pourazarm, "Optimal Routing and Energy Allocation for Lifetime Maximization of Wireless Sensor Networks with Nonideal Batteries." *IEEE Transactions on Control of Network Systems*, Vol. 1, No. 1, 2014, pp. 86–98.

[10] Yetgin, H., K. T. K. Cheung, M. El-Hajjar, and L. Hanzo, "Network-Lifetime Maximization of Wireless Sensor Networks," *IEEE Access*, Vol. 3, 2015, pp. 2191–2226.

[11] Zhang H., and J. C. Hou, "On the Upper Bound of α-Lifetime for Large Sensor Networks," *ACM Transactions on Sensor Networks (TOSN)*, Vol. 1, No. 2, 2005, pp. 272–300.

[12] Wu, K., Y. Gao, F. Li, and Y. Xiao, "Lightweight Deployment-Aware Scheduling for Wireless Sensor Networks," *Mobile Networks and Applications*, Vol. 10, No. 6, 2005, pp. 837–852.

[13] Ben Hamida, E., and G. Chelius, "Strategies for Data Dissemination to Mobile Sinks in Wireless Sensor Networks," *IEEE Wireless Communications*, Vol. 15, No. 6, 2008, pp. 31–37.

[14] Yetgin, H., K. T. K. Cheung, M. El-Hajjar, and L. Hanzo Hanzo, "A Survey of Network Lifetime Maximization Techniques in Wireless Sensor Networks," *IEEE Communications Surveys & Tutorials*, Vol. 19, No. 2, 2017, pp. 828–854.

[15] Chang, C.-Y., J.-P. Sheu, Y.-C. Chen, and S.-W. Chang, "An Obstacle-Free and Power-Efficient Deployment Algorithm for Wireless Sensor Networks," *IEEE Transactions on Systems, Man, and Cybernetics-Part A: Systems and Humans*, Vol. 39, No. 4, 2009, pp. 795–806.

[16] Marcelloni, F., and M. Vecchio, "A Simple Algorithm for Data Compression in Wireless Sensor Networks," *IEEE Communications Letters*, Vol. 12, No. 6, 2008.

[17] Yang, S., H.Cheng, and F. Wang, "Genetic Algorithms with Immigrants and Memory Schemes for Dynamic Shortest Path Routing Problems in Mobile Ad Hoc Networks,"

IEEE Transactions on Systems, Man, and Cybernetics, Part C (Applications and Reviews), Vol. 40, No. 1, 2010, pp. 52–63.

[18] Okdem, S., and D. Karaboga, "Routing in Wireless Sensor Networks Using an Ant Colony Optimization (ACO) Router Chip," *Sensors*, Vol. 9, No. 2, 2009, pp. 909–921.

[19] Liang, Y., J. Cao, L. Zhang, R. Wang, and Q. Pan, "A Biologically Inspired Sensor Wakeup Control Method for Wireless Sensor Networks," *IEEE Transactions on Systems, Man, and Cybernetics, Part C (Applications and Reviews)*, Vol. 40, No. 5, 2010, pp. 525–538.

[20] Park, T. R., K.-J. Park, and M. J. Lee, "Design and Analysis of Asynchronous Wakeup for Wireless Sensor Networks," *IEEE Transactions on Wireless Communications*, Vol. 8, No. 11, 2009.

[21] Hu, X.-M., J. Zhang, Y. Yu, et al., "Hybrid Genetic Algorithm Using a Forward Encoding Scheme for Lifetime Maximization of Wireless Sensor Networks," *IEEE Transactions on Evolutionary Computation*, Vol. 14, No. 5, 2010, pp. 766–781.

[22] Lin, Y., J. Zhang, H. S.-H. Chung, W. H. Ip, Y. Li, and Y.-H. Shi, "An Ant Colony Optimization Approach for Maximizing the Lifetime of Heterogeneous Wireless Sensor Networks," *IEEE Transactions on Systems, Man, and Cybernetics, Part C (Applications and Reviews)*, Vol. 42, No. 3, 2012, pp. 408–420.

[23] Mendes, L. D. P., and J. J. P. C. Rodrigues, "A Survey on Cross-Layer Solutions for Wireless Sensor Networks," *Journal of Network and Computer Applications*, Vol. 34, No. 2, 2011, pp. 523–534.

[24] Cui, S., R. Madan, A. J. Goldsmith, and S. Lall, "Crosslayer Energy and Delay Optimization in Small-Scale Sensor Networks," *IEEE Transactions on Wireless Communications*, Vol. 6, No. 10, 2007.

[25] Shi, L., and A. O. Fapojuwo, "TDMA Scheduling with Optimized Energy Efficiency and Minimum Delay in Clustered Wireless Sensor Networks," *IEEE Transactions on Mobile Computing*, Vol. 9, No. 7, 2010, pp. 927–940.

[26] Madan, R., S. Cui, S. Lall, and A. Goldsmith, "Crosslayer Design for Lifetime Maximization in Interference-Limited Wireless Sensor Networks," *IEEE Transactions on Wireless Communications*, Vol. 5, No. 11, 2006.

[27] Kwon, H., T. H. Kim, S. Choi, and B. G. Lee, "A Cross-Layer Strategy for Energy-Efficient Reliable Delivery in Wireless Sensor Networks," *IEEE Transactions on Wireless Communications*, Vol. 5, No. 12, 2006.

[28] Ehsan, S., B. Hamdaoui, and M. Guizani, "Radio and Medium Access Contention Aware Routing for Lifetime Maximization in Multichannel Sensor Networks," *IEEE Transactions on Wireless Communications*, Vol. 11, No. 9, 2012, pp. 3058–3067.

[29] Jeon, J.-H., H.-J. Byun, and J.-T. Lim. "Joint Contention and Sleep Control for Lifetime Maximization in Wireless Sensor Networks," *IEEE Communications Letters*, Vol. 17, No. 2, 2013, pp. 269–272.

[30] Sun, Y., S. Bhadra, S.-H. Choi, M. Fu, and X. Lu, "Achieving Robust Sensor Networks Through Close Coordination between Narrow-Band Power Line and Low-Power Wireless Communications," in *Proc. ACM Conference on Embedded Networked Sensor Systems*, Seattle, WA, 2011, pp.373–374.

[31] Tonello, A. M., S. D'Alessandro, F. Versolatto, and C. Tornelli, "Comparison of Narrow-Band OFDM PLC solutions and I-UWB Modulation over Distribution Grids," in *Proc. IEEE International Conference on Smart Grid Communications (SmartGridComm)*, Brussels, Belgium, 2011, pp. 149–154.

[32] Gassara, H., F. Rouissi, and A. Ghazel, "Statistical Characterization of the Indoor Low-Voltage Narrowband Power Line Communication Channel," *IEEE Transactions on Electromagnetic Compatibility*, Vol. 56, No. 1, 2014, pp. 123–131.

[33] Goldsmith, A. J., and S.-G. Chua, "Variable-Rate Variable-Power MQAM for Fading Channels," *IEEE Transactions on Communications*, 45, No. 10, 1997, pp. 1218–1230.

[34] Kazimierczuk, M. K., *RF Power Amplifiers*, Chichester, United Kingdom: John Wiley & Sons, 2014.

[35] Cui, S., A. J. Goldsmith, A. Bahai, et al., "Energy-Constrained Modulation Optimization," *IEEE transactions on Wireless Communications*, Vol. 4, No. 5, 2005, pp. 2349–2360.

[36] Boyd S., and L. Vandenberghe, *Convex Optimization*, Cambridge, United Kingdom: Cambridge university press, 2004.

[37] Wang, H., Y. Yang, M. Ma, J. He, and X. Wang, "Network Lifetime Maximization with Cross-Layer Design in Wireless Sensor Networks," *IEEE Transactions on Wireless Communications*, Vol. 7, No. 10, 2008, pp. 3759–3768.

[38] IEEE Standards Association, IEEE Standard for Broadband over Power Line Networks: Medium Access Control and Physical Layer Specifications, IEEE Std 1901, 2010, 2010, pp. 1–1586.

[39] Yetgin, H., K. T. K. Cheung, M. El-Hajjar, and L.Hanzo, "Cross-Layer Network Lifetime Optimisation Considering Transmit and Signal Processing Power in Wireless Sensor Networks," *IET Wireless Sensor Systems*, Vol. 4, No. 4, 2014, pp. 176–182.

[40] Grant, M., S. Boyd, and Y.Ye, CVX: MATLAB Software for Disciplined Convex Programming, 2008.

[41] Varga, A., Omnet++, "Modeling and Tools for Network Simulation," Springer, Berlin, Heidelberg, 2010, pp. 35–59.

[42] Kellerbauer, H., and H. Hirsch, "Simulation of Powerline Communication with OMNeT++ and INET-Framework," in *Proc. IEEE International Symposium on Power Line Communications and Its Applications (ISPLC)*, Udine, Italy, 2011, pp. 213–217.

[43] Marrocco, G., D. Statovci, and S. Trautmann, "A PLC Broadband Channel Simulator for Indoor Communications," in *Proc. IEEE International Symposium on Power Line Communications and Its Applications (ISPLC)*, Johannesburg, South Africa, 2013, pp. 321–326.

[44] Tonello, A. M., and F. Versolatto, "Bottom-Up Statistical PLC Channel Modeling, Part I: Random Topology Model and Efficient Transfer Function Computation," *IEEE Transactions on Power Delivery*, Vol. 26, No. 2, April 201, pp. 891–898.

10

Hybrid Wireless-Power Line Video Sensor Networks with Distributed Cross-Layer Optimization

Wireless video sensor networks (WVSNs), which are one of the high data rate sensors (or multimedia sensor networks), are tasked to perform video capturing and processing and to deliver the processed video content to a remote control unit (or the sink node) via wireless channels for further information analysis and decision making [1]. Unlike standard mains-powered WVSNs [2], battery-operated WVSNs may find widespread applications in fields such as impromptu surveillance installation and indoor elder care and home security due to the advantages of discreet and unobtrusive installation and removal. In addition, battery-operated WVSNs are immune to failure of the power distribution system. Several battery-powered wireless cameras used for home security, baby monitoring, and assisted living are being manufactured and marketed, such as Arlo [3], and Circle designed by Logitech [4].

In typical scenarios, battery-powered WVSNs are supposed to support high data rates and provide high-quality video, which necessitate huge power consumption at the video sensor. Although battery replacement may be feasible in certain scenarios, replacing batteries in a large number of video sensors regularly is cumbersome. Consequently, maintaining energy consumption at a low level is critical for WVSNs. This chapter therefore focuses on improving the network lifetime through HVSN, which includes both hybrid wireless and PL video sensor nodes.

10.1 Related Work

In [5], a distributed algorithm for maximizing the network lifetime of WVSNs is proposed based on the power-rate-distortion (P-R-D) model [6]. However, the channel capacity is assumed to be unlimited. The authors in [7] studied the optimization trade-off between network lifetime and video distortion by jointly considering source/channel rate adaptation and network coding for a energy-constrained WVSNs. A distributed algorithm is proposed in [8] to achieve optimal trade-off between network lifetime and video distortion by joint design of coding and routing optimization in WVSNs with correlated sources. In [9], the authors studied the placement design of motion sensors and cameras in order to maximize the network lifetime, in which the cameras are activated whenever motion is detected.

Rate/channel adaptation has been proven as an effective means of enhancing the wireless network efficiency [10,11]. In channel adaptation, the data being communicated is considered to be generic and therefore generally encoded at the source with fixed rates [7]. However, it is difficult to fully utilize network resources with flexible channel adaptation together with a predetermined source rate. When the instantaneous channel capacity fails to meet the predetermined source rate, network congestion would occur and could never be prevented by any rate adaptation scheme. On the contrary, the channel would be underutilized. In [12], the authors investigated the nature of source data and proposed adaptive source encoding rates to satisfy the distortion constraints. However, the system they studied is a single-hop wireless system. In this work, a joint source/channel rate adaptation framework for multihop multipath video sensor network will be studied.

10.2 Chapter Overview

In this chapter, an HVSN that consists of both battery-powered wireless sensor nodes and PL sensor nodes to is proposed to maximize the network lifetime. This work is different in the following aspects. First, to the best of our knowledge, it is the first reported work to investigate video sensor networks with hybrid power sources and hybrid communication schemes. The proposed HVSN takes advantage of the flexibility of wireless sensors while PL sensors are deployed to prolong the network lifetime. Second, the joint design of video encoding rate, aggregate power consumption, and channel access control, along with link rate allocation, is studied to maximize the hybrid network lifetime. The joint design achieves much better performance than separate optimization. Third, a distributed algorithm for the network lifetime maximization problem is proposed. The proposed distributed algorithm shares the computational burden among all nodes with much lower communication overhead. Fourth, the impact of dynamic network change and network scalability is studied.

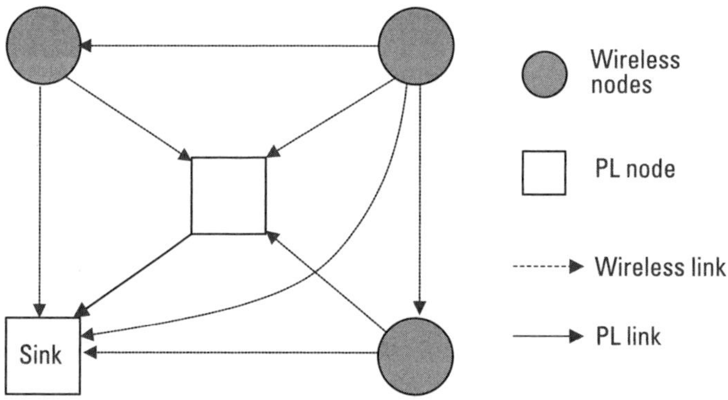

Figure 10.1 Topology of an example HVSN.

The rest of this chapter is organized as follows. Section 10.3 describes the system model. Section 10.4 formulates the optimization problem. In Section 10.5, a distributed algorithm is proposed. Section 10.6 analyzes the numerical results. This chapter is summarized in Section 10.7.

10.3 System Model

An HVSN that includes both wireless video sensor nodes and PL sensor nodes is studied, as depicted in Figure 10.1. The wireless video sensor nodes are placed high above the room and perform video capturing, encoding, and routing. The PL sensor nodes simply perform as relay nodes to help to forward the video content collected by the wireless nodes to the sink node, which is the remote control unit acting as destinations of the HVSN. The PL nodes are assumed to be mounted with wireless receivers so that they can receive wireless signals.

Notations: The wireless nodes set is denoted by $\mathbb{W} = \{1, ..., |W|\}$ and the PL nodes set is indicated by $\mathbb{P} = \{1, ..., |P|\}$. S is used to denote the single sink node in the network. L_w is defined as the number of wireless links and L_p is the number of PL links. $\mathcal{L}_w$ and $\mathcal{L}_p$ are used to label all the wireless and PL links so that $\mathcal{L}_w = \{1, ..., L_w\}$ and $\mathcal{L}_p = \{1, ..., L_p\}$. $l_w \in \mathcal{L}_w$ is the index used to denote the lth wireless link and $l_p \in \mathcal{L}_p$ is used to represent the lth PL link. $\mathcal{L} = \mathcal{L}_w \bigcup \mathcal{L}_p$ denotes all the links in the HVSN and $l \in \mathcal{L}$ indicates the lth link, which is from a transmitter node i to a receiver node j and is denoted by (i, j). Finally, $\mathcal{O}(i)$ and $\mathcal{I}(i)$ represent the set of outgoing and incoming links at node i, respectively.

10.3.1 Video Distortion Model

Unlike traditional WSNs, the video content captured by the video sensor network is first compressed locally before being injected into the channel for transmission.

In video communication over lossy channels, the end-to-end distortion D is divided into two parts: (1) source coding distortion D_c caused by video compression, and (2) transmission distortion D_t owing to channel errors. Since the encoding and transmission errors are generally uncorrelated [13], it follows that

$$D = D_c + D_t \tag{10.1}$$

in mean squared error (MSE). This model is widely used to estimate the end-to-end distortion in the literature [6,7,14]. For the distortion caused by video compression, an analytic P-R-D model is established in [6], which relates the encoding rate $R^{(i)}$, power consumption due to video encoding $P_c^{(i)}$, and the distortion caused by video compression $D_c^{(i)}$ for each wireless node i as

$$D_c^{(i)} = \sigma^2 e^{-\gamma \cdot R^{(i)} \cdot \left(P_c^{(i)}\right)^{2/3}} \tag{10.2}$$

where γ is a factor related to the encoding efficiency, and σ^2 represents the average input variance. Here, the distortion caused by video compression $D_c^{(i)}$ is defined as the distortion of the video quality by comparing the video content after compression to the original one, and is often measured by MSE.

Figure 10.2 shows the encoding distortion D_c, measured in MSE, as a function of source rate after compression, R and encoding power P_c. Apparently, a target encoding distortion is achievable by adjusting either the encoding power or the source rate. If the encoding power P_c or the source rate after compression R is decreased, then the encoding distortion D_c increases for lack of enough power

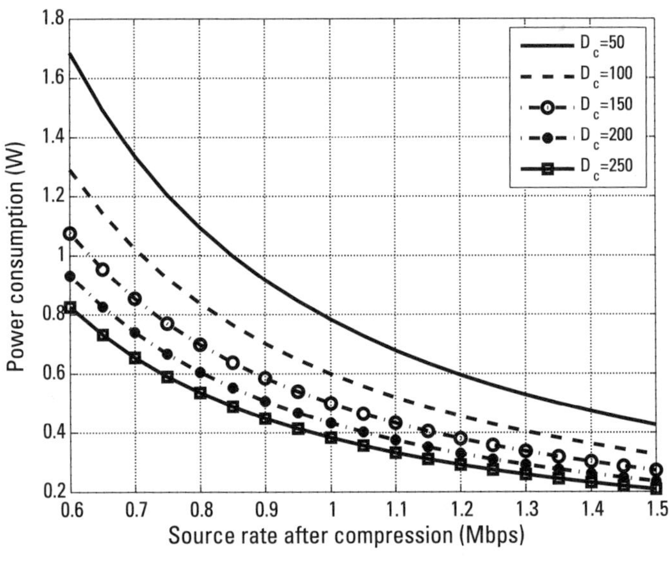

Figure 10.2 Relationship of video encoding power, compressed source rate, and distortion.

for compression. On the other hand, with the increase of encoding power P_c, the transmission power will consequently decrease, which also results in the increase of the distortion D_c. Therefore, an allocation of R and P_c should be balanced to save the power consumption and adjust the video distortion.

Since the source rate shown in Figure 10.2 is the source rate after compression. Assuming that the source rate of the original video content is 5 Mbps and the encoding distortion is fixed, in the first case, the video content is compressed to the source rate R of 2 Mbps, and in the second case, the video content is compressed to the source rate R of 1 Mbps. Apparently, compressing the original video content from 5 Mbps to 1 Mbps would consume more power as compared to compressing the original video content from 5 Mbps to 2 Mbps. Therefore, less power consumption is needed for a higher source rate after compression.

On the other hand, for the distortion caused by transmission, it is shown in [15] that after a certain threshold of BER is achieved, the video quality would not increase significantly with further decrease in BER. This indicates that although channel conditions are rather variable, the distortion can be neglected with a proper target BER value.

10.3.2 Channel Access Model

A widely adopted MAC protocol in sensor networks is the contention-based MAC protocol [16,17]. In this chapter, the p-persistent contention based MAC protocol is used. In such a protocol [16,17], each node i has a certain persistence probability P_i to compete for channel access. It is assumed that time is split into intervals and the transmission of the node begins at the start of each interval. If node i is ready for transmission, it picks a link $l \in \mathcal{O}(i)$ (i.e., $(i,j) \in \mathcal{O}(i)$) out of all its outgoing wireless or PL links with probability q_l, and competes to access the channel with persistence probability P_i. Hence, link $l \in \mathcal{O}(i)$ has a transmission attempt probability $p_l = q_l \cdot P_i$, where $\sum_{l \in \mathcal{O}(i)} q_l = 1$. Therefore, the persistence probability is

$$P_i = \sum_{l \in \mathcal{O}(i)} p_l \tag{10.3}$$

where $0 \le p_l \le 1, \forall l \in \mathcal{L}$, and $0 \le P_i \le 1, \forall i \in \mathbb{W} \bigcup \mathbb{P}$.

In continuous video acquiring applications, assuming the packet loss probability through link l is ε_l, the success probability for packet transmission can be expressed as

$$\tau_l = (1 - \varepsilon_l) \cdot p_l \cdot \prod_{k \in N_l^I} (1 - P_k) \tag{10.4}$$

where N_l^I is the set of nodes whose transmissions introduces interferences to the end node of link l. For wireless links, it is assumed that any outgoing link of node

m interferes with link (i,j) if $d_{(m,j)} < (1+\Phi)\cdot d_{(i,j)}$, where $\Phi \geq 0$ represents the range of interference. The average throughput of link l can thus be defined as

$$c_l = C_l^0 \cdot \tau_l \tag{10.5}$$

where C_l^0 is the maximum rate support by the channel at link l. In addition, the information flow rate f_l on link l is limited by the link capacity, as

$$f_l \leq c_l, \forall l \in \mathcal{L} \tag{10.6}$$

In order to obtain the maximum transmission rate C_l^0, the ITU indoor path loss model [18] is used for the wireless links, as

$$G_l = 20\cdot \log_{10}(f) + 10\cdot n\cdot \log_{10}(d_l) + Lf(n) - 28dB \tag{10.7}$$

where f is the transmission frequency in MHz, n is the path loss exponent, d_l is the transmission range in m and $Lf(n) = 0$ for same floor transmission. For PL links, the random PLC channel generator [19] is used to determine the channel gain, G_l.

MQAM is assumed in this chapter as well as assuming the noise in both links is additive white Gaussian noise with power spectral density N_l. Then, with the corresponding transmission power $\bar{p}_l$ and transmission bandwidth B_l, the instantaneous transmission rate is determined as [20]

$$C_l^0 = B_l\cdot log_2(1 + \frac{K\cdot G_l\cdot \bar{p}_l}{N_l\cdot B_l}) \tag{10.8}$$

where $K = -1.5/\ln(5 \cdot BER)$ is the maximum possible coding gain given a target BER, BER, for modulation schemes such as MQAM [21].

10.3.3 Flow Conservation Constraint

In this chapter, each wireless node is tasked to capture and compress video that should be delivered to a single destination. Then the video traffic is generated with a source rate $R^{(i)}$ in each node i, which can be obtained from (10.2). Note that $R^{(i)} = 0$ for all PL nodes that is, $\forall i \in \mathbb{P}$, since the PL nodes perform as relay nodes. For the sink node, the source rate is defined as $R^{(S)} = -\sum_{i\in\mathbb{W}} R^{(i)}$. Therefore, for each node i, the following constraint holds;

$$\sum\nolimits_{l\in\mathcal{I}(i)} f_l + R^{(i)} = \sum\nolimits_{l\in\mathcal{O}(i)} f_l, \forall i \in \mathbb{W} \bigcup \mathbb{P} \tag{10.9}$$

where f_l is the information flow rate on link l. The flow conservation law simply states that for each node, the outgoing information flow rate should be equal to the incoming information flow rate plus the data rate generated locally.

10.3.4 Energy Consumption Model

In the HVSN, the energy consumption of the wireless nodes are of interest since the battery capacity of these nodes limits the network lifetime. In this chapter, the total power consumption of a wireless node is caused by video encoding, data transmission, and reception.

The power consumption due to video compression can be calculated by the P-R-D model, as in (10.2). According to the power consumption model widely adopted in WVSNs [7,8], the power consumption caused by transmission at wireless node i is expressed as

$$P_t^{(i)} = \sum\nolimits_{l \in \mathcal{O}(i)} (\alpha + \beta \cdot d_l^n) \cdot \frac{f_l}{\tau_l} \tag{10.10}$$

where f_l is the rate assigned on link l, τ_l is the probability for a successful packet transmission of link l, and $\frac{f_l}{\tau_l}$ is the actual rate transmitted through link l. α denotes the energy cost of the transmit electronics, β represents a coefficient relating to the energy cost of the transmit amplifier, d_l is the transmission range of link l, and n is the path-loss exponent [20].

The data reception power consumption at node i is

$$P_r^{(i)} = c^r \cdot \sum\nolimits_{l \in \mathcal{I}(i)} \frac{f_l}{\tau_l} \tag{10.11}$$

where c^r is the energy consumption cost of the radio receiver and $\sum_{l \in \mathcal{I}(i)} \frac{f_l}{\tau_l}$ is the actual aggregate rate transmitted to node i.

Therefore, the overall power consumption at wireless node i can be expressed as

$$P^{(i)} = \left[\frac{1}{\gamma \cdot R^{(i)}} \cdot \ln\left(\frac{\sigma^2}{D_c^{(i)}} \right) \right]^{\frac{3}{2}} + P_t^{(i)} + P_r^{(i)} \tag{10.12}$$

10.3.5 Network Lifetime

The network lifetime is considered as the duration from the beginning of the network until the first wireless node running out of energy. In the HVSN, the battery capacity of each node $i \in \mathbb{W}$ is denoted as $E^{(i)}$. Therefore, the lifetime of each node i is

$$T_i = \frac{E^{(i)}}{P^{(i)}}, \forall i \in \mathbb{W} \tag{10.13}$$

Hence, the network lifetime is

$$T_{net} = \min_{i \in \mathbb{W}} \{T_i\} = \min_{i \in \mathbb{W}} \{\frac{E^{(i)}}{P^{(i)}}\} \tag{10.14}$$

10.4 Problem Formulation

The objective of the problem under study is to maximize the network lifetime in a predetermined static HVSN topology with the instantaneous transmission rate on each link and the initial battery capacity of each wireless node. The joint optimization decisions are the video encoding rate, the encoding power, and the routing path as well as the channel contention resolution on each link. In addition, a predefined video quality should be satisfied.

$$\mathbf{P1}: \quad \max_{(\mathbf{f},\mathbf{R},\mathbf{p},\mathbf{T})} \quad [\min_{i\in\mathbb{W}}\{T_i\}] \tag{10.15}$$

$$\text{s.t.} \quad \sigma^2 e^{-\gamma\cdot R^{(i)}\cdot\left(P_c^{(i)}\right)^{2/3}} \le D_c^{(i)}, \quad \forall i \in \mathbb{W} \tag{10.16}$$

$$\frac{E^{(i)}}{T_i} = P_c^{(i)} + \sum\nolimits_{l\in\mathcal{O}(i)} (\alpha + \beta\cdot d_l^n)\cdot\frac{f_l}{\tau_l} + c^r\cdot\sum\nolimits_{l\in\mathcal{I}(i)}\frac{f_l}{\tau_l}, \quad \forall i \in \mathbb{W} \tag{10.17}$$

$$0 \le p_l \le 1, \quad \forall l \in \mathcal{L} \tag{10.18}$$

$$0 \le P_i \le 1, \quad \forall i \in \mathbb{W}\bigcup\mathbb{P} \tag{10.19}$$

$$f_l \ge 0, \quad \forall l \in \mathcal{L} \tag{10.20}$$

$$R^{(i)} \ge 0, \quad \forall i \in \mathbb{W} \tag{10.21}$$

along with (10.3)–(10.6) and (10.9). Constraint (10.16) represents that the encoding distortion should not exceed the corresponding upper bound on each wireless node. Constraint (10.17) reflects the power consumption of each wireless video sensor node. By observation, variables $P_c^{(i)}$, τ_l and P_i are dummy variables since these can be determined in expressions of other variables. Hence, the optimization variables in **P1** are f_l, $R^{(i)}$, p_l, and T_i.

10.5 Optimization Approach and Distributed Algorithm

The problem in P1 is not convex due to nonlinearity in constraints (10.4) and (10.17). In order to convert the problem to a convex problem, constraint (10.4) can be reformulated by taking the logarithm on both sides. Also, variable $q_i = 1/T_i$ is introduced in (10.17) as node i's normalized power consumption in relation to its battery capacity $E^{(i)}$. Hence, the objective function becomes

$$\max(\min_{i\in\mathbb{W}}\{T_i\}) = \min(\max_{i\in\mathbb{W}}\{q_i\})$$

In addition, (10.16) is simplified by taking the logarithm on both sides. Further, $F_l = f_l/\tau_l$ is introduced as the total aggregated data flow rate on each link.

The objective function, $\max_{i \in \mathbb{W}}\{q_i\}$, however, is nondifferentiable and needs all sensor nodes' information. Hence, it is difficult to develop a fully distributed algorithm to solve the problem. One solution is to introduce $q_i = q_j$ as the constraints, and the objective function is equivalent to $\sum_{i \in \mathbb{W}} q_i^2$.

Therefore, the optimization problem in P1 becomes

$$\mathbf{P2}: \quad \min_{(\mathbf{F},\mathbf{R},\mathbf{p},\mathbf{q})} \sum_{i \in \mathbb{W}} q_i^2 \tag{10.22}$$

$$\text{s.t.} \quad \sum\nolimits_{l \in \mathcal{O}(i)} F_l \cdot \tau_l - \sum\nolimits_{l \in \mathcal{I}(i)} F_l \cdot \tau_l = R^{(i)}, \forall i \in \mathbb{W} \bigcup \mathbb{P} \tag{10.23}$$

$$\frac{1}{\gamma \cdot \left(P_c^{(i)}\right)^{2/3}} \cdot \ln\left(\frac{\sigma^2}{D_c^{(i)}}\right) \le R^{(i)}, \quad \forall i \in \mathbb{W} \tag{10.24}$$

$$F_l \le C_l^0, \quad \forall l \in \mathcal{L} \tag{10.25}$$

$$E^{(i)} \cdot q_i = P_c^{(i)} + \sum\nolimits_{l \in \mathcal{O}(i)} (\alpha + \beta \cdot d_l^n) \cdot F_l + c^r \cdot \sum\nolimits_{l \in \mathcal{I}(i)} F_l, \quad \forall i \in \mathbb{W} \tag{10.26}$$

$$\ln \tau_l = \ln(1 - \varepsilon_l) \cdot p_l + \sum\nolimits_{k \in N_l^I} \ln(1 - P_k), \quad \forall l \in \mathcal{L} \tag{10.27}$$

$$P_i = \sum\nolimits_{l \in \mathcal{O}(i)} p_l, \quad \forall l \in \mathcal{L} \tag{10.28}$$

$$F_l \ge 0, \quad \forall l \in \mathcal{L} \tag{10.29}$$

$$0 \le p_l \le 1, \quad \forall l \in \mathcal{L} \tag{10.30}$$

$$0 \le P_i \le 1, \quad \forall i \in \mathbb{W} \bigcup \mathbb{P} \tag{10.31}$$

$$R^{(i)} \ge 0, \quad \forall i \in \mathbb{W} \tag{10.32}$$

$$q_i = q_j, \quad \forall i, j \in \mathbb{W} \tag{10.33}$$

The optimization variables are $F_l, R^{(i)}$, p_l, and q_i. In P2, it can be proved that the objective function is strictly convex, the equality constraints are affine, and the inequality constraints are convex. Hence, P2 is a convex optimization problem [22].

To develop a distributed algorithm for P2, the primal decomposition method [23] is used regarding the coupling variable F_l, which results in a two-level optimization problem. At the lower level, there is

$$\text{P2-a:} \quad \min_{(\mathbf{R},\mathbf{p},\mathbf{q})} \sum_{i \in \mathbb{W}} q_i^2 \tag{10.34}$$

$$s.t.\quad \sum\nolimits_{l\in\mathcal{O}(i)} F_l \cdot \tau_l - \sum\nolimits_{l\in\mathcal{I}(i)} F_l \cdot \tau_l = R^{(i)}, \quad \forall i \in \mathbb{W} \bigcup \mathbb{P} \tag{10.35}$$

$$\frac{1}{\gamma \cdot \left(P_c^{(i)}\right)^{2/3}} \cdot \ln\left(\frac{\sigma^2}{D_c^{(i)}}\right) \le R^{(i)}, \quad \forall i \in \mathbb{W} \tag{10.36}$$

$$E^{(i)} \cdot q_i = P_c^{(i)} + \sum\nolimits_{l\in\mathcal{O}(i)} (\alpha + \beta \cdot d_l^n) \cdot F_l + c^r \cdot \sum\nolimits_{l\in\mathcal{I}(i)} F_l \tag{10.37}$$

$$\ln \tau_l = \ln(1-\varepsilon_l)\cdot p_l + \sum\nolimits_{k\in N_l^I} \ln(1-P_k), \quad \forall l \in \mathcal{L} \tag{10.38}$$

$$P_i = \sum\nolimits_{l\in\mathcal{O}(i)} p_l, \quad \forall l \in \mathcal{L} \tag{10.39}$$

$$0 \le p_l \le 1, \quad \forall l \in \mathcal{L} \tag{10.40}$$

$$0 \le P_i \le 1, \quad \forall i \in \mathbb{W} \bigcup \mathbb{P} \tag{10.41}$$

$$R^{(i)} \ge 0, \quad \forall i \in \mathbb{W} \tag{10.42}$$

$$q_i = q_j, \quad \forall i,j \in \mathbb{W} \tag{10.43}$$

and at the higher level, it gives

$$\text{P2-b:} \quad \min_{(\mathbf{F})} \quad U^*(\mathbf{F}) \tag{10.44}$$

$$s.t.\quad F_l \le C_l^0, \quad \forall l \in \mathcal{L} \tag{10.45}$$

$$F_l \ge 0, \quad \forall l \in \mathcal{L} \tag{10.46}$$

P2-a performs a low-level optimization when the coupling variable $F_l = f_l/\tau_l$ is fixed, while P2-b performs a high-level optimization to update F_l. $U^*(F)$ is the optimal value of the objective function in P2-a for given variables, F_l. The output of the low-level optimization is locally optimal and provides an approximation to the global optimal solution using the result of the high-level optimization.

10.5.1 Low-Level Optimization

To solve the low-level optimization in problem P2-a, the constraints (10.35), (10.36), and (10.43) in P2-a are relaxed, which yield the following Lagrangian [22]:

$$\mathbf{L}(\lambda, \theta, \nu, \mathbf{R}, \mathbf{p}, \mathbf{q}) = \sum_{i\in\mathbb{W}} q_i^2 + \sum_{i\in\mathbb{W}\bigcup\mathbb{P}} \lambda_i \cdot \Big(\sum\nolimits_{l\in\mathcal{O}(i)} F_l \cdot \tau_l - \sum\nolimits_{l\in\mathcal{I}(i)} F_l \cdot \tau_l - R^{(i)}\Big)$$

$$+\sum_{i\in\mathbb{W}}\theta_i\cdot[\frac{1}{\gamma\cdot(P_c^{(i)})^{2/3}}\cdot\ln(\frac{\sigma^2}{D_c^{(i)}})-R^{(i)}]$$
$$+\sum_{l\in\mathcal{L}_w}\nu_l\cdot(q_i-q_j) \quad (10.47)$$

where λ, θ, and ν are the Lagrange multipliers corresponding to constraints (10.35), (10.36), and (10.43), respectively. In addition, the corresponding Lagrange dual function is

$$g(\lambda,\theta,\nu)=\inf_{(\mathbf{R},\mathbf{p},\mathbf{q})}\mathbf{L}(\lambda,\theta,\nu,\mathbf{R},\mathbf{p},\mathbf{q})$$
$$s.t.\quad \text{Constraints (10.37) - (10.42) in}\quad \mathbf{P2-a} \quad (10.48)$$

The Lagrange dual problem of P2-a is then defined as

$$\mathbf{P2-a-1:}\quad \max\quad g(\lambda,\theta,\nu)$$
$$s.t.\quad \theta_i\geq 0,\quad \forall i\in\mathbb{W} \quad (10.49)$$

The corresponding Lagrange multipliers can be solved with the subgradient method as

$$\lambda_i(n_L+1)=\lambda_i(n_L)+\omega(n_L)\cdot(\sum\nolimits_{l\in\mathcal{O}(i)}F_l\cdot\tau_l-\sum\nolimits_{l\in\mathcal{I}(i)}F_l\cdot\tau_l-R^{(i)}) \quad (10.50)$$

$$\theta_i(n_L+1)=\{\theta_i(n_L)+\omega(n_L)\cdot[\frac{1}{\gamma\cdot(P_c^{(i)})^{2/3}}\cdot\ln(\frac{\sigma^2}{D_c^{(i)}})-R^{(i)}]\}^+ \quad (10.51)$$

$$\nu_l(n_L+1)=\nu_l(n_L)+\omega(n_L)\cdot(q_{\mathcal{O}^{-1}(l)}-q_{\mathcal{I}^{-1}(l)}) \quad (10.52)$$

where n_L represents the low-level iteration index, $\{\cdot\}^+$ denotes the projection onto the set of nonnegative real numbers, and $\omega(n_L)$ is a positive step size in low-level optimization problems. $\mathcal{I}^{-1}(l)$ denotes the node associated with the incoming link and l, $\mathcal{O}^{-1}(l)$ denotes the node associated with the outgoing link l.

According to convex optimization theorem [23], if the original problem P2-a is convex, it is equivalent to its Lagrange dual problem in (10.49). Then, the low-level optimization problem P2-a can be further decomposed into a set of subproblems P2-a-2 to P2-a-4 that can be solved in a distributed manner,

$$\mathbf{P2-a-2:}\quad \min\quad \sum_{i\in\mathbb{W}}q_i^2+\sum_{i\in\mathbb{W}}\theta_i\cdot[\frac{1}{\gamma\cdot(P_c^{(i)})^{2/3}}\cdot\ln(\frac{\sigma^2}{D_c^{(i)}})]$$
$$+\sum_{l\in\mathcal{L}_w}\nu_l\cdot(q_i-q_j)$$
$$s.t.\quad (10.37)$$

$$\mathbf{P2-a-3:}\quad \min \quad -\sum_{i\in\mathbb{W}}\lambda_i\cdot R^{(i)}-\sum_{i\in\mathbb{W}}\theta_i\cdot R^{(i)}$$
$$s.t.\quad (10.42)$$

$$\mathbf{P2-a-4:}\quad \min \quad \sum_{i\in\mathbb{W}\bigcup\mathbb{P}}\lambda_i\cdot(\sum\nolimits_{l\in\mathcal{O}(i)}F_l\cdot\tau_l-\sum\nolimits_{l\in\mathcal{I}(i)}F_l\cdot\tau_l)$$
$$s.t.\quad (10.38)-(10.41)$$

Subproblem P2-a-2 is the energy conservation in wireless sensor nodes taking into account impacts both from the MAC layer and the network layer. Subproblem P2-a-3 is the source rate control problem at the application layer. Subproblem P2-a-4 is the channel contention resolution problem at the MAC layer. These three problems are solved separately and coordinated by Lagrange multipliers λ, θ and ν.

Energy conservation problem (application layer): The variable q_i in P2-a-2 can be solved using the subgradient algorithm as

$$\begin{aligned}q_i(n_L+1)&=\{q_i(n_L)-\omega(n_L)\cdot\frac{\partial\mathbf{L}(\lambda,\theta,\nu,\mathbf{R},\mathbf{p},\mathbf{q})}{\partial q_i}\}^+\\&=\{q_i(n_L)-\omega(n_L)\cdot[2\cdot q_i-\frac{2}{3}\cdot\theta_i\cdot\frac{\ln(\frac{\sigma^2}{D_c^{(i)}})}{\gamma}\cdot(P_c^{(i)})^{-5/3}\cdot E^{(i)}\\&\quad+(\sum\nolimits_{l\in\mathcal{O}(i)}\nu_l-\sum\nolimits_{l\in\mathcal{I}(i)}\nu_l)]\}^+\end{aligned}\tag{10.53}$$

where $P_c^{(i)}$ can be obtained as

$$P_c^{(i)}=E^{(i)}\cdot q_i-\sum\nolimits_{l\in\mathcal{O}(i)}(\alpha+\beta\cdot d_l^n)\cdot F_l-c^r\cdot\sum\nolimits_{l\in\mathcal{I}(i)}F_l\tag{10.54}$$

The energy conservation at each wireless node i is achieved by adjusting the value of F_l and q_i, with θ_i working as the energy consumption price, and ν_l as the energy balancing price.

Source rate control problem (application layer): The variable $R^{(i)}$ can be updated using the subgradient algorithm as

$$\begin{aligned}R_i(n_L+1)&=\{R_i(n_L)-\omega(n_L)\cdot\frac{\partial\mathbf{L}(\lambda,\theta,\nu,\mathbf{R},\mathbf{p},\mathbf{q})}{\partial R_i}\}^+\\&=\{R_i(n_L)-\omega(n_L)\cdot(-\lambda_i-\theta_i)\}^+\end{aligned}\tag{10.55}$$

Channel contention resolution problem (MAC layer): The channel contention resolution problem aims to find the optimal transmission persistence probabilities of the links under given variable F_l and price λ_i. Its problem formulation is similar to that in [16]. Hence, the same algorithm can be used in which defining

$$\mu_i = \sum\nolimits_{l \in \mathcal{O}(i)} F_l \cdot (\lambda_{\mathcal{I}^{-1}(l)} - \lambda_{\mathcal{O}^{-1}(l)}) + \sum\nolimits_{k \in L^{\mathcal{I}}(i)} F_k \cdot (\lambda_{\mathcal{I}^{-1}(k)} - \lambda_{\mathcal{O}^{-1}(k)}) \tag{10.56}$$

where $L^{\mathcal{I}}(i)$ represents the set of links whose transmission get interfered from the transmission of node i, excluding outgoing links from node i. Then the transmission attempt probability of the link l is

$$p_l = \begin{cases} \frac{F_l \cdot (\lambda_{\mathcal{I}^{-1}(l)} - \lambda_{\mathcal{O}^{-1}(l)})}{\mu_i} & , \quad \mu_i \neq 0 \\ \frac{1}{|\mathcal{O}(i)| + |L^{\mathcal{I}}(i)|} & , \quad \mu_i = 0 \end{cases} \tag{10.57}$$

10.5.2 High-Level Optimization

The high-level optimization problem aims to find the routing and link rate allocation, which is in the network layer. Suppose $\hat{\tau}_l$, $\hat{\lambda}_{O^{-1}(l)}$ and $\hat{\lambda}_{I^{-1}(l)}$ are the optimal variable and Lagrange price corresponding to (10.35) in problem P2-a. The optimization approach proceeds as taking a dual decomposition with respect to (10.45) in problem P2-b and thus formulates the Lagrangian as

$$\mathrm{L}'(\varphi, \mathrm{F}) = U^*(\mathrm{F}) + \sum_{l \in \mathcal{L}} \varphi_l (F_l - C_l^0) \tag{10.58}$$

where φ is the Lagrangian multiplier. The optimal value of F_l can be found by the subgradient algorithm

$$\begin{aligned} F_l(n_H + 1) &= \{F_l(n_H) - \varpi(n_H) \cdot \frac{\partial \mathrm{L}'(\varphi, \mathrm{F})}{\partial F_l}\}^+ \\ &= \{F_l(n_H) - \varpi(n_H) \cdot (\hat{\tau}_l \cdot (\hat{\lambda}_{\mathcal{O}^{-1}(l)} - \hat{\lambda}_{\mathcal{I}^{-1}(l)}) + \varphi_l)\}^+ \end{aligned} \tag{10.59}$$

and the corresponding Lagrangian dual variable is updated as

$$\varphi_l(n_H + 1) = \{\varphi_l(n_H) + \varpi(n_H) \cdot (F_l - C_l^0)\}^+ \tag{10.60}$$

where n_H denotes the high-level iteration index, and $\varpi(n_H)$ is the positive step size of the high-level optimization problem.

10.5.3 Summary of the Distributed Algorithm

The distributed implementation of the proposed two-level iterative algorithm is summarized in Algorithm 1. In Algorithm 1, each link $l = (i, j)$ is delegated to its sender node i, and all computations related to that link will be executed on node i. It can be seen that the communication overhead at each iteration consists

Algorithm 1 Distributed Two-Level Optimization Algorithm

Initialization • Set $n_L = 1$ and $n_H = 1$ • Initialize optimization variables q_i, $R^{(i)}$, θ_i, $\forall i \in \mathbb{W}$, and λ_i, $\forall i \in \mathbb{W} \bigcup \mathbb{P}$, ν_i, $\forall i \in \mathbb{W}$ and p_l, F_l, φ_l, $\forall l \in \mathcal{L}$ with, e.g., zeros. **I. Low-level implementation a.** Update at each wireless node, $\forall i \in \mathbb{W}$ 1. Update $\theta_i(n_L)$, $\nu_l(n_L)$, $q_i(n_L)$ and $R^{(i)}(n_L)$ according to (10.51), (10.52), (10.53) and (10.55), respectively. 2. Communicate the updated dual variable $\nu_i(n_L+1)$ to the end nodes of incoming links $l \in \mathcal{I}(i)$. **b.** Update at each wireless and PL node, $\forall i \in \mathbb{W} \bigcup \mathbb{P}$ 1. Update $\lambda_i(n_L)$ according to (10.50). 2. Communicate the updated dual variable $\lambda_i(n_L+1)$ to the end nodes of incoming links $l \in \mathcal{I}(i)$. **c.** Update at each wireless and PL link, $\forall l \in \mathcal{L}$ 1. Update $p_l(n_L)$ according to (10.57). **II. High-level implementation a.** Update at each wireless and PL link, $\forall l \in \mathcal{L}$ 1. Update $F_l(n_H)$ and $\varphi_l(n_H)$ according to (10.59) and (10.60), respectively. 2. Communicate the updated actual transmission rate $F_l(n_H+1)$ to the corresponding end nodes of outgoing links $l \in \mathcal{O}(i)$. All variables converge to the optimums.

of conveying $\lambda_i(n_L+1)$, $\nu_i(n_L+1)$, and $F_l(n_H+1)$ to the corresponding nodes. Thus compared to the main stream of video transmission traffic, the communication overhead introduced by such information exchange is quite small.

The proposed distributed algorithm needs to be implemented whenever the initial network starts monitoring or the dynamic change of the network condition suddenly happens, in order to catch up with the optimal network lifetime for the network.

10.6 Numerical Results

In this section, the overall performance of the proposed distributed algorithm will be evaluated. The topology used for the HVSN is a square area of 20m × 20m where eight nodes are randomly placed (including the sink node, as shown in Figure 10.3). Without loss of generality, the node located at position (0, 0) is considered to be the sink node. For performance comparison, two WVSN topologies are considered: (1) the same as HVSN except that all nodes are wireless nodes and (2) the same as the topology in (1) except that PL nodes are removed. In the following, the topologies in (1) and (2) are referred to as 8-node and 6-node WVSNs, respectively. Numerically, the values of all the related model parameters are listed in Table 10.1. Also, the upper bound of the encoding distortion $D_c^{(i)}$ in MSE is set to 100 if not specified otherwise. For PL links, the random PLC channel generator [19] is used to determine the channel gain. BB PLC with a total bandwidth of 26 MHz consisting 917 subcarriers is considered. Each PL node is allocated with five subcarriers. The transmission power of the PL link is

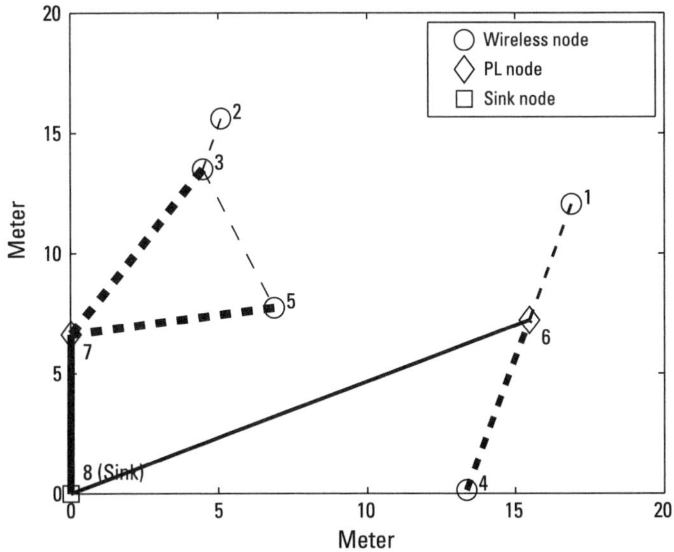

Figure 10.3 Randomly generated topology of HVSN with illustration of aggregate link rates (blue line indicates a wireless link, red line represents a PL link, the thickness of the line is proportional to the aggregate link rate).

Table 10.1
Configuration of Model Parameters in the HVSN

Parameter	Description	Value
σ^2	Average input variance of the video in MSE	3500
γ	Encoding efficiency coefficient	5 $W^{3/2}/Mb/s$
α	Energy cost of the transmit electronics	0.2 J/Mb
β	Coefficient term of the transmit amplifier	1.3×10^{-8} $J/Mb/m^4$
c^r	Energy consumption cost of the radio receiver	0.1 J/Mb
$E^{(i)}$	The initial energy at wireless node i	2 MJ
f_w	Radio frequency	900 MHz
$\overline{P}_{l_w}$	Transmission power of wireless link	0.5 W
$\overline{B}_{l_w}$	Transmission bandwidth of wireless link	1 MHz
N_{l_w}	Noise PSD level of wireless link	−131 dBm/Hz
BER	Target BER	10^{-4}
n	Wireless path loss exponent	4
ε_l	Packet loss rate at link l	0.1

−50 dBm/Hz and the noise PSD level is −120 dBm/Hz. A fixed step size of 0.01 is used.

Figures 10.4, 10.5, and 10.6 show the convergence behavior of the proposed distributed algorithm with the illustrations of iterations of node lifetime, source

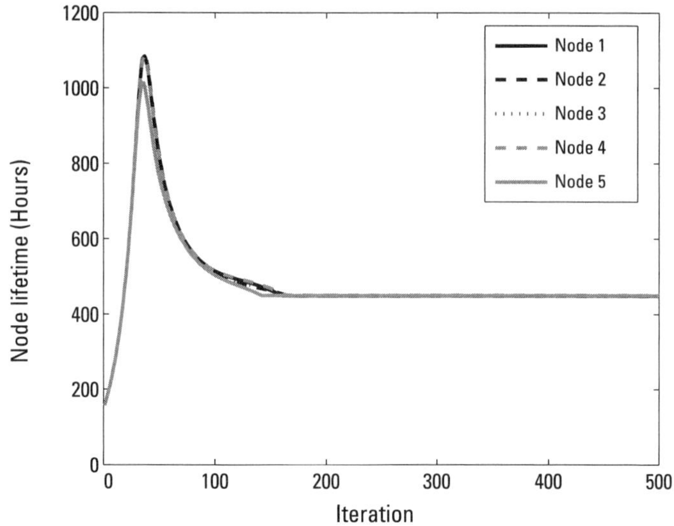

Figure 10.4 Iterations of node lifetime.

rate, and link transmission attempt probabilities, respectively, based on the topology shown in Figure 10.3. It can be observed that these variables converge to the optimum value within 220 iterations. Therefore, the proposed algorithm can converge to the steady state in a relatively short period of time. Also, Figure 10.4 depicts that the node lifetime of each node in the network can achieve a common node lifetime in the steady state. The proposed distributed algorithm allows each node to exchange its local information about node lifetime with its neighboring nodes and thus can effectively balance the energy consumption among all nodes.

Figure 10.3 shows the aggregate link rates, where the thickness of the line is proportional to the aggregate link rate. Combining Figures 10.3 and 10.5, it is found that the nodes with heavy duty in relaying the data tend to encode the video with a lower encoding power (and hence larger source rate), thus saving the power for data transmission and reception. For example, node 3 relays the traffic originating from node 2, and therefore node 3 consumes much less energy for video encoding (consequently, node 3 has a much larger source rate than node 2), which saves energy for data transmission and reception. However, node 2 depends on other nodes to transmit its data to the sink node, so it consumes much energy for video encoding (node 2 has a much lower source rate), which saves the transmission and reception energy for the relay nodes.

Another example of the randomly generated topology is shown in Figure 10.7, where the topology is a square area of 40m × 40m where 12 nodes (including the sink node), which includes eight wireless nodes, are randomly placed.

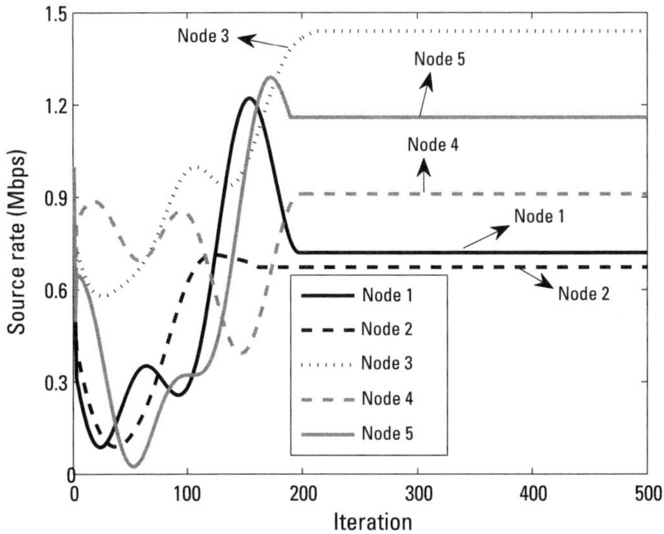

Figure 10.5 Iterations of source rate.

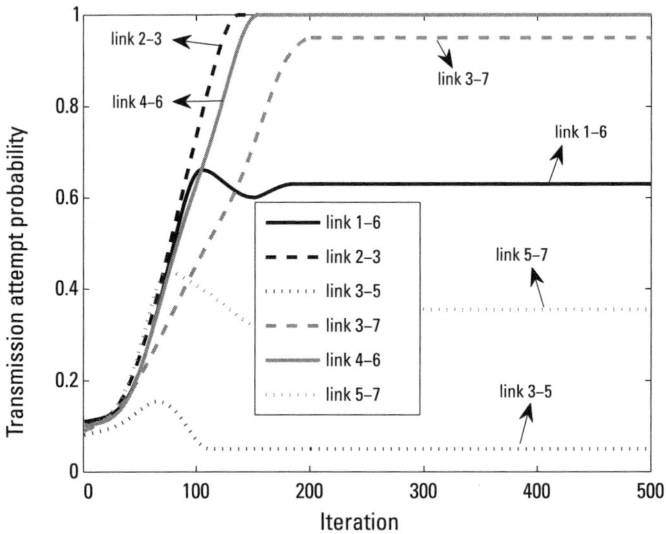

Figure 10.6 Iterations of link transmission attempt probability.

Figures 10.8, 10.9, and 10.10 show the convergence behavior of the proposed distributed algorithm with the illustrations of iterations of node lifetime, source rate, and link transmission attempting probabilities, respectively, based on the topology shown in Figure 10.7. It can be observed that these variables converge to the optimum value within 300 iterations. Therefore, the proposed algorithm can converge to the steady state in a relatively short period of time.

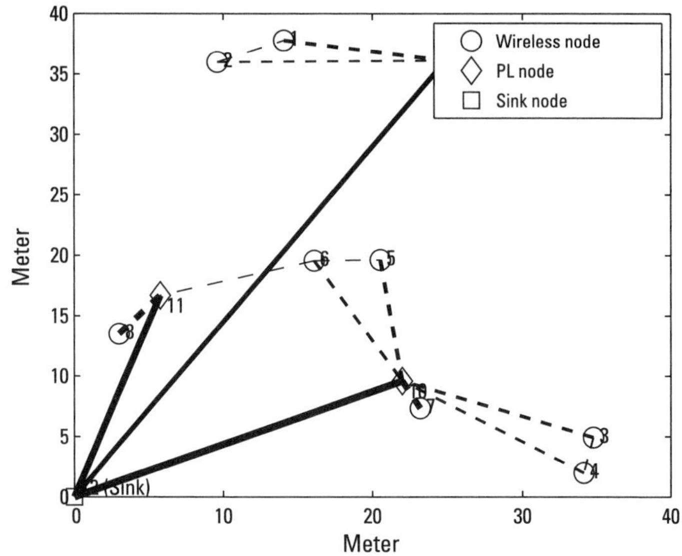

Figure 10.7 Randomly generated topology of HVSN with illustration of aggregate link rates (blue line indicates a wireless link, red line represents a PL link, the thickness of the line is proportional to the aggregate link rate).

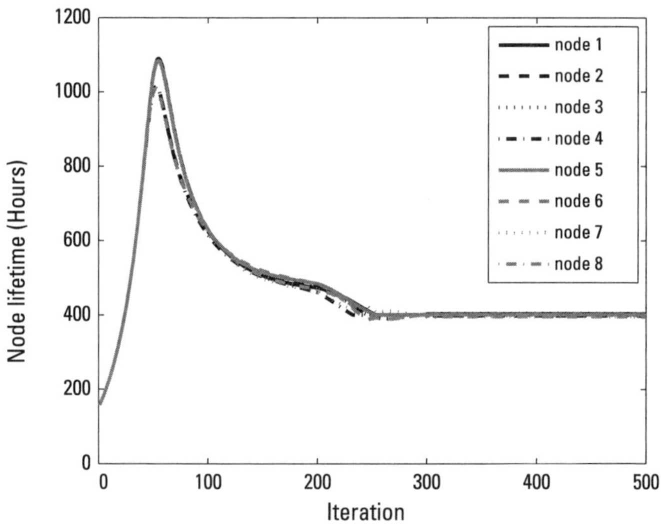

Figure 10.8 Iterations of node lifetime.

Compared to the convergence speed of the algorithm on the network topology shown in Figure 10.3, it is clear that more iterations are required for a more complex network topology. As well, Figure 10.8 depicts that the node lifetime of each node in the network can achieve a common node lifetime in the steady state.

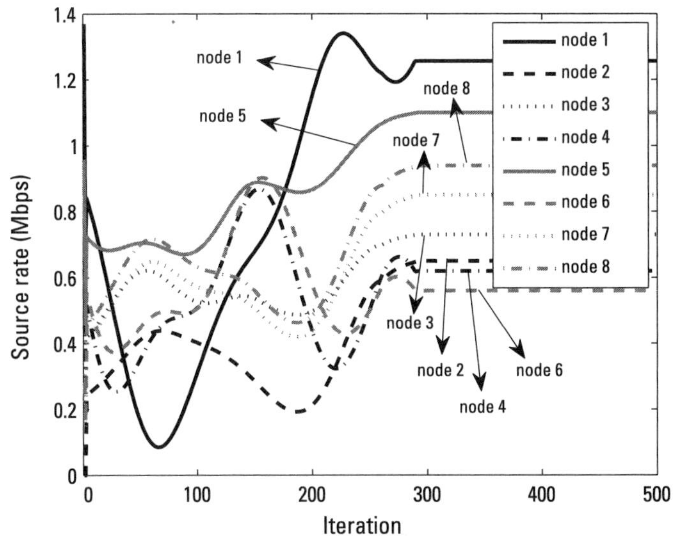

Figure 10.9 Iterations of source rate.

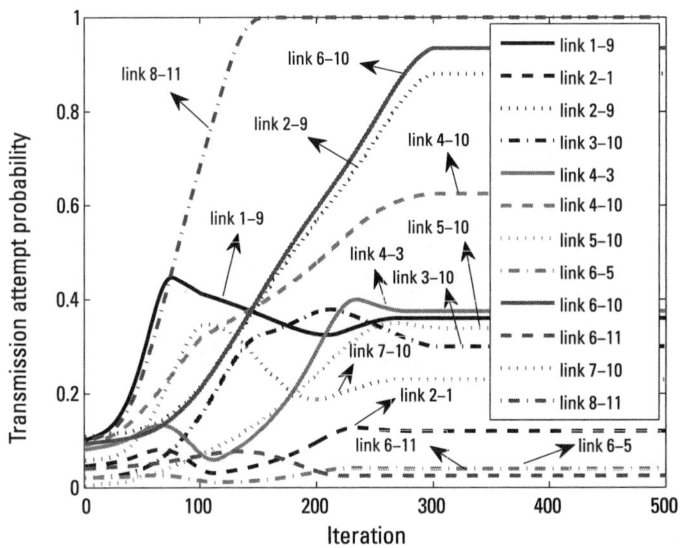

Figure 10.10 Iterations of link transmission attempt probability.

Figure 10.11 depicts the network lifetime of the proposed HVSN (shown in Figure 10.3), the 6-node WVSN and the 8-node WVSN under different distortion requirements. With a smaller distortion requirement, the network lifetime of the networks decreases. In this case, each node either consumes more power for video encoding, or encodes the video content with a larger source rate, which increases the transmission and reception power consumption of each

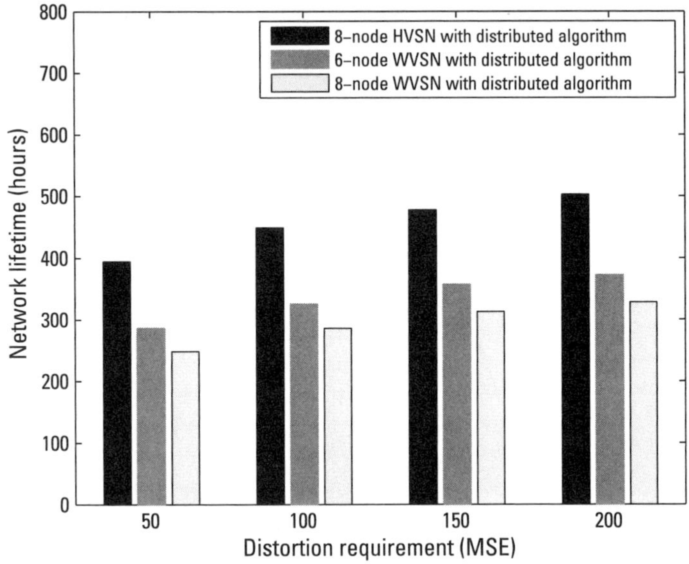

Figure 10.11 Comparison of the network lifetime under different distortion requirements.

node. By comparing the network lifetime of the proposed 8-node HVSN with the pure WVSN (*i.e.*, the 6-node WVSN and the 8-node WVSN) it can be deduced that the network lifetime is increased by around 37.5% and 58.1%, respectively, under different distortion requirements. The deployment of PL nodes reduces the channel access contention for the wireless links. This is also the reason that 6-node WVSN has a longer network lifetime than the 8-node WVSN, as with the increase of wireless nodes, each wireless link has a lower transmission attempt probability, and thus consumes more power for data transmission.

The convergence behaviour of the proposed distributed algorithm under dynamic changes of the video content for the HVSN illustrated in Figure 10.3 is shown in Figure 10.12. The video content is characterized through the average input variance σ^2. At the startup of the distributed algorithm, the average input variance is set to be 3,500. After it reaches the steady state, at iteration 300, the average input variance is changed to 2,500. It can be observed that the maximum power dissipation among all nodes adapt themselves to this change and reach another steady state in around 140 iterations. It can also be observed that the steady-state maximum power dissipation when the average input variance is changed to 2,500 has a smaller value than that when the average input variance is 3,500. Then at iteration 600, the average input variance is changed from 2,500 to 5,500. Again, the algorithm quickly adapts itself to converge to a new steady state in around 140 iterations. It can be observed that the steady-state maximum power dissipation when the average input variance is changed to 5,500 has a larger value than that when the average input variance is 3,500.

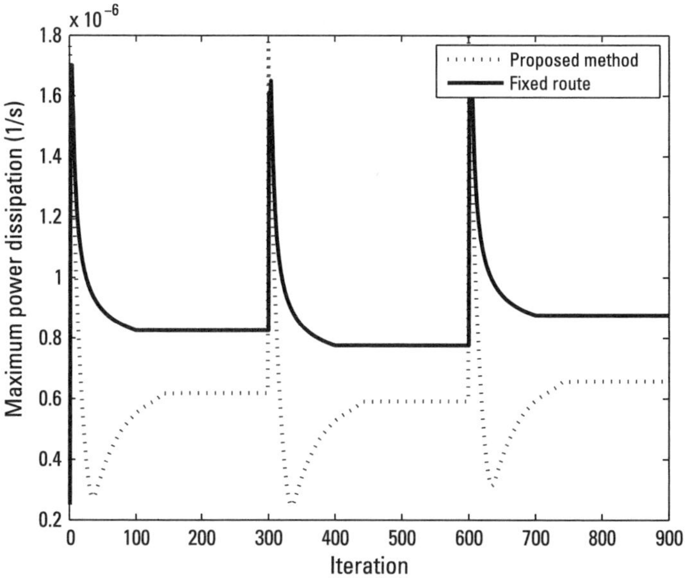

Figure 10.12 Convergence behaviour of the proposed algorithms with dynamic changes of the video content.

Figure 10.12 compares the convergence speed and performance of the proposed algorithm with the fixed route algorithm. In the fixed route algorithm, each node has a selection of predefined paths to the sink node, which is generated by existing routing algorithms such as Dijkstra's algorithm, while other optimization decisions are the same as in the proposed algorithm. Other optimization parameters are the same as in the proposed algorithm. It can be observed that the algorithm with a fixed route converges faster (within 100 iterations) than the proposed algorithm with respect to the change of the average input variance. However, since a selection of predefined paths to the sink node is used for each node, this may result in some battery-powered wireless nodes becoming the hot spots where other nodes may relay data to. Consequently, these nodes will be drained out of energy very rapidly, and hence the algorithm with fixed routes has a much shorter network lifetime.

Figure 10.13 shows the convergence behavior of the proposed distributed algorithm under dynamic changes of the the network topology for the HVSN shown in Figure 10.3. Initially, the HVSN with eight nodes reaches a steady state. At iteration 300, node 5 (shown in Figure 10.3) is turned off. The algorithm detects the topology change, and adapts itself to the change by reaching a new steady state in around 140 iterations. This actually causes the network lifetime to increase significantly since with the removal of node 5, other wireless nodes can access the wireless channel for a longer time. Later at iteration 600, node 5 is

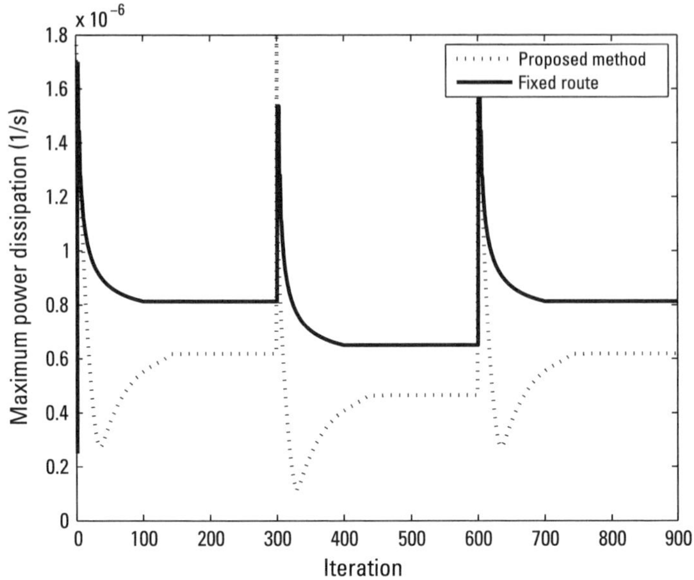

Figure 10.13 Convergence behavior of the proposed algorithms with dynamic changes of the network topology.

turned back on. The algorithm automatically adapts to the topology change and converges to a new steady state in around 140 iterations.

Again, Figure 10.13 compares the convergence speed and performance of the proposed algorithm with the fixed route algorithm. It can be observed that the algorithm with a fixed route generally converges within 100 iterations, as compared to 140 iterations by the proposed algorithm with respect to the removal and addition of node 5. Again, since a selection of predefined paths to the sink node is used for each node, the routing decision for each node is not optimized. Therefore, this may result in some battery-powered wireless nodes becoming the hot spots where other nodes may relay data to, or the node may send data directly to the sink node other than the PL nodes. Consequently, these nodes will be drained out of energy very rapidly, and hence the algorithm with fixed routes has a much shorter network lifetime.

From the analysis of Figures 10.12 and 10.13, it demonstrates that the proposed distributed algorithm can quickly adapt itself under dynamic changes of the video content or network topology. This indicates that the proposed algorithm is suitable for the application where frequent sensor node repositioning is required. In addition, by comparing the proposed algorithm to the algorithm with fixed routes for each node, it demonstrates the effectiveness of the proposed cross-layer optimization method.

In order to assess the impact of network scale on the performance of the proposed HVSN, three larger network topologies are considered. In the first case,

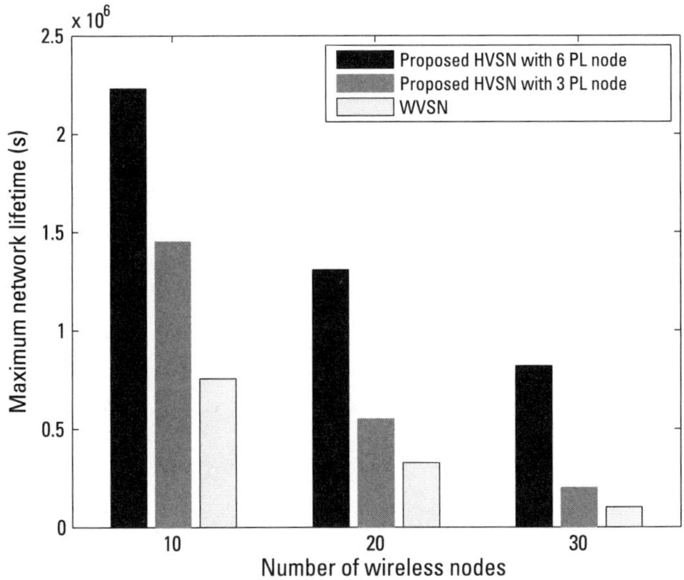

Figure 10.14 Comparison of network lifetime under different network scales.

10 nodes are randomly scattered in an area of 40m × 40m. In the second case, 20 nodes are randomly deployed in an area of 60m × 60m. In the third case, 30 nodes are placed in an area of 80m × 80m in a random fashion. The results are averaged over 50 random realizations in each case for fair comparison. For the topology setup of the HVSN, three PL nodes and six PL nodes are randomly placed on the network in each case. Figure 10.14 shows that with the increase of network scale, the network lifetime of the 6-node HVSN, 3-node HVSN, and pure WVSN decreases rapidly. This is due to the fact that with more wireless nodes in the network, more video content is generated that needs to be forwarded to the sink node, which increases the power dissipation at each node. Also, with the increase of the number of wireless links, each link has a lower transmission attempt probability, which further increases the power consumption. By comparing the network lifetime of HVSN and WVSN, it is clear that the HVSN can improve network lifetime significantly, especially when the number of PL nodes is large in the network. However, as in the case of a network with 30 nodes, the 3-node HVSN improves the network lifetime by a small amount. This is due to the fact that the throughput of PL links cannot support the amount of data generated in the network.

10.7 Summary

In this chapter, a distributed algorithm for network lifetime maximization in HVSN was proposed by joint design of video encoding rate, aggregate power consumption, and channel access control, along with link rate allocation. Each node solves the optimization problem locally and only requires information exchange with its neighbouring nodes. Through numerical simulations, the performance of the proposed algorithm is evaluated. It is shown that the proposed HVSN can improve the network lifetime of a pure WVSN significantly (by around 37.5% and 58.1%, with 8-node HVSN compared with the 6-node WVSN and 8-node WVSN). The proposed algorithm also adapts rapidly with dynamic changes of video content and network topology. By comparing the proposed algorithm with the algorithm with fixed routes for each node, it was observed that the proposed algorithm has a longer convergence time (140 iterations compared to 100 iterations) while it can achieve a much longer network lifetime. This proves the effectiveness of the proposed cross-layer algorithm.

References

[1] He, Y., L. Guan, and W. Zhu, *Optimal Resource Allocation for Distributed Video Communication*, Boca Raton, FL: CRC Press, 2013.

[2] Gasparini, L., R. Manduchi, M. Gottardi, and D. Petri, "An Ultralow-Power Wireless Camera Node: Development and Performance Analysis," *IEEE Transactions on Instrumentation and Measurement*, Vol. 60, No. 12, 2011 pp. 3824–3832.

[3] https://www.arlo.com/.

[4] http://www.logitech.com/en-gb/product/circle/.

[5] He, Y.,Ivan Lee, and L. Guan, "Distributed Algorithms for Network Lifetime Maximization in Wireless Visual Sensor Networks," *IEEE Transactions on Circuits and Systems for Video Technology*, Vol. 19, No. 5, 2009, pp. 704–718.

[6] He, Z., and D. Wu, "Resource Allocation and Performance Analysis of Wireless Video Sensors," *IEEE transactions on Circuits and Systems for Video Technology*, Vol. 16, No. 5, 2006, pp. 590–599.

[7] Zou, J., H. Xiong, C. Li, R. Zhang, and Z. He, "Lifetime and Distortion Optimization with Joint Source/Channel Rate Adaptation and Network Coding-Based Error Control in Wireless Video Sensor Networks," *IEEE Transactions on Vehicular Technology*, 60, No. 3, 2011, pp. 1182–1194.

[8] Li, C., J. Zou, H. Xiong, and C. W. Chen, "Joint Coding/ Routing Optimization for Distributed Video Sources in Wireless Visual Sensor Networks," *IEEE Transactions on Circuits and Systems for Video Technology*, Vol. 21, No. 2, 2011, pp. 141–155.

[9] Yang, N., I. Demirkol, and W.Heinzelman, "Motion Sensor and Camera Placement Design for In-Home Wireless Video Monitoring Systems," in *Proc. IEEE Global Telecommunications Conference (GLOBECOM 2011),*Houston, Texas, 2011, pp.1–5.

[10] Chung, S. T., and A. J. Goldsmith, "Degrees of Freedom in Adaptive Modulation: A Unified View," *IEEE Transactions on Communications*, Vol. 49, No. 9, 2001, pp. 1561–1571.

[11] Djonin, D. V., A. K. Karmokar, and V. K. Bhargava, "Joint Rate and Power Adaptation for Type-I Hybrid ARQ Systems Over Correlated Fading Channels Under Different Buffer-Cost Constraints," *IEEE Transactions on Vehicular Technology*, Vol. 57, No. 1, 2008, pp. 421–435.

[12] Li, J. C. F., S. Dey, and J. Evans, "Maximal Lifetime Power and Rate Allocation for Wireless Sensor Systems with Data Distortion Constraints," *IEEE Transactions on Signal Processing*, Vol. 56, No. 5, 2008, pp. 2076–2090.

[13] Stuhlmuller, K., N. Farber, M. Link, and B. Girod, "Analysis of Video Transmission Over Lossy Channels," *IEEE Journal on Selected Areas in Communications*, Vol. 18, No. 6, 2000, pp. 1012–1032.

[14] Zhu, X., E. Setton, and B. Girod, "Congestion–Distortion Optimized Video Transmission Over Ad Hoc Networks," *Signal Processing: Image Communication*, Vol. 20, No. 8, 2005, pp. 773–783.

[15] Pudlewski, S., N. Cen, Z. Guan, and T. Melodia, "Video Transmission Over Lossy Wireless Networks: A Cross-Layer Perspective," *IEEE Journal of Selected Topics in Signal Processing*, Vol. 9, No. 1, 2015, pp. 6–21.

[16] Lee, J.-W., M. Chiang, and R. A. Calderbank, "Optimal MAC Design Based on Utility Maximization: Reverse and Forward Engineering," in *Proc. IEEE International Conference on Computer Communications (INFOCOM)*, Barcelona, Spain, 2006, pp. 1–13.

[17] Zhu, J., S. Chen, B. Bensaou, and K.- L. Hung, "Tradeoff between Lifetime and Rate Allocation in Wireless Sensor Networks: A Cross Layer Approach," in *Proc. IEEE International Conference on Computer Communications (INFOCOM)*, Anchorage, Alaska, 2007, pp. 267–275.

[18] Seybold, J. S., *Introduction to RF Propagation*, Hoboken, NJ: John Wiley & Sons, 2005.

[19] http://www.plc.uma.es/channels.htm/.

[20] Goldsmith, A., *Wireless Communications*, Cambridge, United Kingdom: Cambridge University Press, 2005.

[21] Goldsmith, A. J., and S.-G. Chua, "Variable-Rate Variable-Power MQAM for Fading Channels," *IEEE Transactions on Communications*, Vol. 45, No. 10, 1997, pp. 1218–1230.

[22] Boyd S., and L. Vandenberghe, *Convex Optimization*, Cambridge, United Kingdom: Cambridge University Press, 2004.

[23] Palomar, D. P., and M. Chiang, "A Tutorial on Decomposition Methods for Network Utility Maximization," *IEEE Journal on Selected Areas in Communications*, Vol. 24, No. 8, 2006, pp. 1439–1451.

Appendix A

Derivation of A General Expression for OBS in Chapter 5

For calculating the *OBS*, there are two cases. First, the branch is connecting a branching node to a terminal load. Second, the branch is connecting two branching nodes. For the first case the *OBS* is represented by geometric series $\frac{a}{1-r}$ where $a = \Gamma_{CB} T_{BC} e^{-2\gamma l_{BC}}$ and $r = \Gamma_{CB} \Gamma_{BC} e^{-2\gamma l_{BC}}$ for the branch BC in Figure 5.1. For the second case, the formula equation that represents the *OBS* gain of the branch connecting two branching nodes is derived. Starting from the formula representing the adjacent nodes gain A_L [4]:

$$A_L = \sum_{l=1}^{k} a_l \tag{A.1}$$

Where k is the total number of the branching nodes on the left-hand side of the Tx, a_l is the effect of the multipath signal propagation resulting from branching node C_l in the lth section and it can be obtained by [4]

$$a_l = [\prod_{p=1}^{l-1} M_p \prod_{q=1}^{l-1} \prod_{s=1}^{n} (OBS_{q,s} + 1)] b_l \tag{A.2}$$

$$M_p = T_{C_p C_{p-1}} T_{C_{p-1} C_p} e^{-2\gamma l_{C_p C_{p-1}}} \tag{A.3}$$

$$b_l = N_l + M_l [\prod_{s=1}^{n} (OBS_{l,s} + 1) - 1] \tag{A.4}$$

$$N_l = \Gamma_{C_l C_{l-1}} T_{C_{l-1} C_l} e^{-2\gamma l_{C_l C_{l-1}}} \tag{A.5}$$

Using A.2, the summation of the two terms a_{k-1} and a_k is given by

$$\begin{aligned} a_{k-1} + a_k = & [\prod_{p=1}^{k-2} M_p \prod_{q=1}^{k-2} \prod_{s=1}^{n} (OBS_{q,s} + 1)] b_{k-1} \\ & + [\prod_{p=1}^{k-1} M_p \prod_{q=1}^{k-1} \prod_{s=1}^{n} (OBS_{q,s} + 1)] b_k \end{aligned} \tag{A.6}$$

Substituting in (A.6) using (A.4)

$$\begin{aligned} a_{k-1} + a_k = & [\prod_{p=1}^{k-2} M_p \prod_{q=1}^{k-2} \prod_{s=1}^{n} (OBS_{q,s} + 1)] b_{k-1} \\ & \times [N_{k-1} + M_{k-1} (\prod_{s=1}^{n} (OBS_{k-1,s} + 1)(b_k + 1) - 1)] \end{aligned} \tag{A.7}$$

This leads us to the conclusion that the b_k is the *OBS* value for the branch that connects both the $(k-1)$-th and the kth branching nodes, hence the number of the branches coming out of the $(k-1)$-th node changes from n to $n+1$ and (A.7) can be rewritten as

$$\begin{aligned} a_{k-1} + a_k = & [\prod_{p=1}^{k-2} M_p \prod_{q=1}^{k-2} \prod_{s=1}^{n} (OBS_{q,s} + 1)] b_{k-1} \\ & \times [N_{k-1} + M_{k-1} (\prod_{s=1}^{n+1} (OBS_{k-1,s} + 1) - 1)] \end{aligned} \tag{A.8}$$

From (A.8), it can be deduced that $OBS_{k-1,s}$ of the sth branch connecting $(k-1)$-th node to the k-th node is equal to b_k. However, if the kth node is a terminal node, then the $OBS_{k-1,s}$ is equal to $\frac{a_s}{1-r_s}$ as shown in (5.3).

Appendix B

Proof of Lemma 1 in Chapter 6

Let $\alpha_{i,k}^{(n)}$ be the ith component of the interference vector α_k. In order to satisfy the interference power limit at the TV PU Rx, while maintaining good SNR at the Rx

$$\Sigma_{i=1}^{n_r} \alpha_{i,k}^{(n)} P_{i,k}^{(n)} \leq \Gamma, \quad \Lambda^{(n)} \mathbf{P}_1^{(n)} \succeq \sigma_n^2 + \sigma_p^2 \tag{B.1}$$

Let $\Gamma \leq \sigma_n^2$ to avoid harmful interference. Also let $\sigma_n^2 + \sigma_p^2 = \rho\Gamma$. Hence, the following equation can be concluded

$$\Lambda^{(n)} \mathbf{P}_1^{(n)} \succeq \rho \alpha_k^{(n)} \mathbf{P}_1^{(n)} \tag{B.2}$$

This can be translated in the following matrix form

$$\begin{bmatrix} \frac{\lambda_1^{(n)}}{\rho} - \alpha_{1,k}^{(n)} & \cdots & -\alpha_{n_r,k}^{(n)} \\ \vdots & & \vdots \\ -\alpha_{1,k}^{(n)} & \cdots & \frac{\lambda_{n_r}^{(n)}}{\rho} - \alpha_{n_r,k}^{(n)} \end{bmatrix} \begin{bmatrix} P_1^{(n)} \\ \vdots \\ P_{n_r}^{(n)} \end{bmatrix} \succeq 0 \tag{B.3}$$

Directly, the condition in (6.7) can be concluded.

Appendix C

Proof of Lemma 2 in Chapter 6

Using the OR decision rule for MIMO channel, it can be deduced

$$\hat{\mathcal{P}}_d = 1 - \Pi_{i=1}^{n_r}(1 - \mathcal{P}_{d_i}) \tag{C.1}$$

where $\mathcal{P}_{d_i}$ can be expressed as follows for the energy detection method [14]:

$$\mathcal{P}_{d_i} = Q\Big(\big(\frac{\epsilon}{\sigma_n^2} - \gamma_i - 1\big)\Big(\sqrt{\frac{\tau_i f_s}{2\gamma_i + 1}}\Big)\Big) \tag{C.2}$$

where ϵ is the detection threshold. N_{min} is assumed to be equal for all the MIMO subchannels. Hence, relationship between the detection probabilities of two MIMO subchannels i and m can be expressed as follows using (6.9):

$$\mathcal{P}_{d_i} = Q\Big(\frac{\gamma_m - \gamma_i}{\gamma_m\sqrt{2\gamma_i + 1}} Q^{-1}(\hat{\mathcal{P}}_{fa}) + Q^{-1}(\mathcal{P}_{d_m})\sqrt{\frac{2\gamma_m + 1}{2\gamma_i + 1}}\Big) \tag{C.3}$$

Substituting (C.3) into (C.1), Lemma 2 can directly be concluded.

About the Authors

Dr. Xu Zhu received a B.Eng. degree with first-class honors from Huazhong University of Science and Technology,Wuhan, China, in 1999, and a Ph.D. degree from the Hong Kong University of Science and Technology, Hong Kong, in 2003. She joined the University of Liverpool in 2003 as an academic member of staff, where she is currently a reader. She also has an academic association with Harbin Institute of Technology, Shenzhen, China. Dr. Zhu has published over 180 peer-reviewed publications by leading international journals and conferences. She has also conducted a high volume of research projects. She has served as an editor for the *IEEE Transactions on Wireless Communications* and a guest editor for several international journals, such as those in the field of electronics. She has been actively engaged in organizing international conferences and workshops as a chair (e.g., Symposium Chair of the IEEE ICC 2016 and 2019, Vice Chair of the 2006 and 2008 ICARN International Workshops, and Publication Chair of the IEEE IUCC-2014 and 2016). Her research interests include MIMO, channel estimation and equalization, millimeter-wave transmission, power line communication, ultra-reliable low-latency communication, green communication, and smart grid communications.

Dr. Kainan Zhu received a B.Eng. degree with first-class honors and a Ph.D. degree in electrical engineering and electronics from the University of Liverpool, United Kingdom, in 2013 and 2019, respectively. In 2013, he was awarded The Institution of Engineering and Technology (IET) Prize 2013. His research interests include wireless sensor networks, wireless video sensor networks, power line communication, and convex optimization.

Dr. Mohammad Heggo received B.Sc. and M.Sc. degrees in electrical engineering from Ain Shams University, Cairo, Egypt, in 2006 and 2010, respectively. In 2007, he joined the Research and Development Department, at Elsewedy Electrometer Company, Egypt, as an embedded communication software engineer. In 2013, he was awarded a dual Ph.D. scholarship for four years from the Electrical Engineering and Electronics Department at the University of Liverpool, United Kingdom, and the Cognitive Communications Technology

Department, Agency for Science, Technology, and Research, Singapore. He was awarded a Ph.D. degree in electrical engineering from the University of Liverpool, United Kingdom, in 2017. He joined the HOME project as a research associate at the Electrical Engineering and Electronics Department, University of Manchester, United Kingdom, from 2017–2019, where he studied high electromagnetic field interference to UAVs that are monitoring offshore windfarms converter stations. Currently, he works as a research associate at the Computing Department, Imperial College of London, United Kingdom, where he shares in the development of new RF machine monitoring technology known as CogniSense. The new technology is predicted to be a revolutionary step change in condition monitoring for industrial machines. He has published 11 papers and presentations in highly reputable and peer-reviewed journals and conferences, spanning different applications of communications and signal processing for future IoT and sensor networks. His research interests include condition monitoring of machines and different electronic devices, signal processing, power line communication, electromagnetic interference to different electrical systems and transmission lines, cognitive radio, MIMO communication systems, and smart grid communications.

Index

Recent Titles in the Artech House Telecommunications and Network Engineering Series

Access Networks: Technology and V5 Interfacing, Alex Gillespie

Achieving Global Information Networking, Eve L. Varma et al.

Advanced High-Frequency Radio Communications, Eric E. Johnson et al.

ATM Interworking in Broadband Wireless Applications, M. Sreetharan and S. Subramaniam

ATM Switches, Edwin R. Coover

ATM Switching Systems, Thomas M. Chen and Stephen S. Liu

Broadband Access Technology, Interfaces, and Management, Alex Gillespie

Broadband Local Loops for High-Speed Internet Access, Maurice Gagnaire

Broadband Networking: ATM, SDH, and SONET, Mike Sexton and Andy Reid

Broadband Telecommunications Technology, Second Edition, Byeong Lee, Minho Kang, and Jonghee Lee

The Business Case for Web-Based Training, Tammy Whalen and David Wright

The Business Privacy Law Handbook, Charles H. Kennedy

Centrex or PBX: The Impact of IP, John R. Abrahams and Mauro Lollo

Chinese Telecommunications Policy, Xu Yan and Douglas Pitt

Communication and Computing for Distributed Multimedia Systems, Guojun Lu

Communications Technology Guide for Business, Richard Downey, Seán Boland, and Phillip Walsh

Community Networks: Lessons from Blacksburg, Virginia, Second Edition, Andrew M. Cohill and Andrea Kavanaugh, editors

Component-Based Network System Engineering, Mark Norris, Rob Davis, and Alan Pengelly

Computer Telephony Integration, Second Edition, Rob Walters

Creating Value-Added Services and Applications for Converged Communications and Networks, Johan Zuidweg

Customer-Centered Telecommunications Services Marketing, Karen G. Strouse

Delay- and Disruption-Tolerant Networking, Stephen Farrell and Vinny Cahill

Deploying and Managing IP over WDM Networks, Joan Serrat and Alex Galis, editors

Desktop Encyclopedia of the Internet, Nathan J. Muller

Digital Clocks for Synchronization and Communications, Masami Kihara, Sadayasu Ono, and Pekka Eskelinen

Digital Modulation Techniques, Second Edition, Fuqin Xiong

Disaster Recovery Planning for Communications and Critical Infrastructure, Leo A. Wrobel and Sharon M. Wrobel

E-Commerce Systems Architecture and Applications, Wasim E. Rajput

EMI Protection for Communication Systems, Kresimir Malaric

Engineering Internet QoS, Sanjay Jha and Mahbub Hassan

Error-Control Block Codes for Communications Engineers, L. H. Charles Lee

Essentials of Modern Telecommunications Systems, Nihal Kularatna and Dileeka Dias

FAX: Facsimile Technology and Systems, Third Edition, Kenneth R. McConnell, Dennis Bodson, and Stephen Urban

Fundamentals of Network Security, John E. Canavan

Gigabit Ethernet Technology and Applications, Mark Norris

The Great Telecom Meltdown, Fred R. Goldstein

Guide to ATM Systems and Technology, Mohammad A. Rahman

A Guide to the TCP/IP Protocol Suite, Floyd Wilder

Home Networking Technologies and Standards, Theodore B. Zahariadis

Hybrid Wireless-Power Line Communication for Indoor IoT Networks, Xu Zhu, Kainan Zhu, and Mohammad Heggo

Implementing Value-Added Telecom Services, Johan Zuidweg

Information Superhighways Revisited: The Economics of Multimedia, Bruce Egan

Installation and Maintenance of SDH/SONET, ATM, xDSL, and Synchronization Networks, José M. Caballero et al.

Integrated Broadband Networks: TCP/IP, ATM, SDH/SONET, and WDM/Optics, Byeong Gi Lee and Woojune Kim

Internet E-mail: Protocols, Standards, and Implementation, Lawrence Hughes

Internet Technologies for Fixed and Mobile Networks, Toni Janevski

Introduction to Communication Networks, Tarmo Anttalainen and Ville Jääskeläinen

Introduction to Telephones and Telephone Systems, Third Edition, A. Michael Noll

An Introduction to U.S. Telecommunications Law, Second Edition, Charles H. Kennedy

IP Convergence: The Next Revolution in Telecommunications, Nathan J. Muller

LANs to WANs: The Complete Management Guide, Nathan J. Muller

The Law and Regulation of Telecommunications Carriers, Henk Brands and Evan T. Leo

Litigating with Electronically Stored Information, Marian K. Riedy, Susman Beros and Kim Sperduto

Managing Internet-Driven Change in International Telecommunications, Rob Frieden

Marketing Telecommunications Services: New Approaches for a Changing Environment, Karen G. Strouse

Mission-Critical Network Planning, Matthew Liotine

Multimedia Communications Networks: Technologies and Services, Mallikarjun Tatipamula and Bhumip Khashnabish, editors

Multimedia Networking Technologies, Protocols, and Architectures, Ivan Vidal, Ignacio Soto, Albert Banchs, Jaime Garcia-Reinoso, Ivan Lozano, and Gonzalo Camarillo

Next Generation Intelligent Networks, Johan Zuidweg

Open Source Software Law, Rod Dixon

Performance Evaluation of Communication Networks, Gary N. Higginbottom

Performance of TCP/IP over ATM Networks, Mahbub Hassan and Mohammed Atiquzzaman

The Physical Layer of Communications Systems, Richard A. Thompson, David Tipper, Prashant Krishnamurthy, and Joseph Kabara

Power Line Communications in Practice, Xavier Carcelle

Practical Guide for Implementing Secure Intranets and Extranets, Kaustubh M. Phaltankar

Practical Internet Law for Business, Kurt M. Saunders

Practical Multiservice LANs: ATM and RF Broadband, Ernest O. Tunmann

Principles of Modern Communications Technology, A. Michael Noll

A Professional's Guide to Data Communication in a TCP/IP World, E. Bryan Carne

Programmable Networks for IP Service Deployment, Alex Galis et al., editors

Protocol Management in Computer Networking, Philippe Byrnes

Pulse Code Modulation Systems Design, William N. Waggener

Reorganizing Data and Voice Networks: Communications Resourcing for Corporate Networks, Thomas R. Koehler

Security, Rights, and Liabilities in E-Commerce, Jeffrey H. Matsuura

Service Assurance for Voice over WiFi and 3G Networks, Richard Lau, Ram Khare, and William Y. Chang

Service Level Management for Enterprise Networks, Lundy Lewis

SIP: Understanding the Session Initiation Protocol, Fourth Edition, Alan B. Johnston

Smart Card Security and Applications, Second Edition, Mike Hendry

SNMP-Based ATM Network Management, Heng Pan

Spectrum Wars: The Policy and Technology Debate, Jennifer A. Manner

Strategic Management in Telecommunications, James K. Shaw

Strategies for Success in the New Telecommunications Marketplace, Karen G. Strouse

Successful Business Strategies Using Telecommunications Services, Martin F. Bartholomew

Teleccomunication Electronics, Dante Del Corso, Vittorio Camarchia, Roberto Quaglia, and Paolo Bardella

Telecommunications Cost Management, S. C. Strother

Telecommunications Department Management, Robert A. Gable

Telecommunications Deregulation and the Information Economy, Second Edition, James K. Shaw

Telecommunications Technology Handbook, Second Edition, Daniel Minoli

Telemetry Systems Engineering, Frank Carden, Russell Jedlicka, and Robert Henry

Telephone Switching Systems, Richard A. Thompson

3D and HD Broadband Video Networking, Benny Bing

Third Networks and Services, Mehmet Toy and Hakki Candan Cankaya

Understanding Modern Telecommunications and the Information Superhighway, John G. Nellist and Elliott M. Gilbert

Understanding Networking Technology: Concepts, Terms, and Trends, Second Edition, Mark Norris

Understanding SIP Servlets 1.1, Chris Boulton and Kristoffer Gronowski

Understanding Voice over IP Security, Alan B. Johnston and David M. Piscitello

Videoconferencing and Videotelephony: Technology and Standards, Second Edition, Richard Schaphorst

Virtualized Software-Defined Networks and Services, Qiang Duan and Mehmet Toy

Visual Telephony, Edward A. Daly and Kathleen J. Hansell

Wide-Area Data Network Performance Engineering, Robert G. Cole and Ravi Ramaswamy

Winning Telco Customers Using Marketing Databases, Rob Mattison

WLANs and WPANs towards 4G Wireless, Ramjee Prasad and Luis Muñoz

World-Class Telecommunications Service Development, Ellen P. Ward

For further information on these and other Artech House titles, including previously considered out-of-print books now available through our In-Print-Forever® (IPF®) program, contact:

Artech House
685 Canton Street
Norwood, MA 02062
Phone: 781-769-9750
Fax: 781-769-6334
e-mail: artech@artechhouse.com

Artech House
16 Sussex Street
London SW1V HRW UK
Phone: +44 (0)20 7596-8750
Fax: +44 (0)20 7630-0166
e-mail: artech-uk@artechhouse.com

Find us on the World Wide Web at: www.artechhouse.com